高等职业教育"十四五"系列教材

机电专业

U0162539

机电工程专业英语

主　编　王　磊
副主编　王叶萍　苏冬云　王　坤
主　审　杨林娟

扫码加入学习圈
轻松解决重难点

南京大学出版社

图书在版编目(CIP)数据

机电工程专业英语 / 王磊主编. —南京：南京大
学出版社，2022.6
ISBN 978 - 7 - 305 - 24135 - 2

Ⅰ.①机… Ⅱ.①王… Ⅲ.①机电工程—英语—教材
Ⅳ.①TH

中国版本图书馆 CIP 数据核字(2020)第 265584 号

出版发行 南京大学出版社
社　　址 南京市汉口路 22 号　　　邮　编　210093
出 版 人 金鑫荣
书　　名 机电工程专业英语
主　　编 王　磊
责任编辑 吕家慧　　　　　　　编辑热线　025 - 83596997
照　　排 南京开卷文化传媒有限公司
印　　刷 南京百花彩色印刷广告制作有限责任公司
开　　本 787×1092 1/16 印张 13 字数 293 千
版　　次 2022 年 6 月第 1 版 2022 年 6 月第 1 次印刷
ISBN 978 - 7 - 305 - 24135 - 2
定　　价 39.00 元

网　　址：http://www.njupco.com
官方微博：http://weibo.com/njupco
微信服务号：njuyuexue
销售咨询热线：(025)83594756

扫码可免费申请
本书教学资源

前 言

本书是根据高等职业教育机电工程类专业英语教育要求编写而成的。本书在编写过程中,充分汲取高等职业教育机电工程专业英语的教学改革经验,以就业为导向、以能力为本位,突出实用性、适应性和针对性,旨在培养学生顺利阅读和使用专业英语的能力,加强综合素质的培养,满足高等职业教育机电工程专业英语教学改革和培养高素质应用型、技能型人才的需要。

本书的选材是在有限的篇幅内尽可能地涵盖机电工程的相关领域,尽量选编新技术和新设备内容。全书精选了22个单元,每个单元包括课文、阅读材料以及相关的词汇表、注释,并附有课文和阅读材料的全篇译文,以便于教师组织教学和学生自学。全书共分为五个部分,每部分有4~6个单元。

第一部分是有关机械、机电工程的基础知识,内容包括材料选用及热处理、机床与加工、焊接、CAD/CAM技术及其应用、计算机图形学等;第二至五部分是有关四个专业领域(机械制造及数控技术、机电一体化技术、模具设计与制造、过程装备技术)的相关知识,内容包括计算机集成制造系统和控制系统、柔性制造系统、计算机数字控制发展史、计算机数控系统和数控机床、工业机器人、机电一体化技术、可编程逻辑控制器、现代控制系统及应用、压力传感器技术、电动机、模具加工技术和成型方法、过程装备基础知识及典型过程设备和典型机器等内容。

全书五部分内容既有内在联系又相对独立,教师可针对具体情况有选择性地安排教学。

本书可作为高等职业教育院校机械制造、数控技术、机电一体化技术、工业机器人技术、模具设计与制造、过程装备技术等专业教学用书，也可作为各类成人院校机械、机电类专业英语教学用书，同时可供机械、机电类相关工程技术人员阅读参考。

本书由南通职业大学王磊任主编，王叶萍、苏冬云、王坤任副主编，参加编写的还有陈淑侠，杨林娟任主审。

本书编写过程中得到了各院校领导和专家同行的大力支持和帮助，南京大学出版社对本书的出版给予了大力支持，在此一并表示衷心的感谢！同时也感谢编者曾经参阅过的大量中外文献和网络资源的作者们。

由于编者水平有限，书中错误和不妥之处在所难免，恳请相关专业领域的专家和广大师生读者批评指正，以利于本书提高完善。

编　者

2022 年 4 月

Contents

Unit 1

The Roles of Engineers in Manufacturing

Many engineers have their function in the designing of products that are to be brought into reality through the processing or fabrication of materials. In this capacity they are a key factor in the material selection-manufacturing procedure. A design engineer, better than any other person, should know what he wants a design to accomplish. He knows what assumptions he has made about service loads and requirements, what service environment the product must withstand, and what appearance he wants the final product to have. In order to meet these requirements he must select and specify the material(s) to be used. In most cases, in order to utilize the material and to enable the product to have the desired form, he knows that certain manufacturing processes will have to be employed. In many instances, the selection of a specific material may dictate what processing must be used. At the same time, when certain processes are to be used, the design may have to be modified in order for the process to be utilized effectively and economically. Certain dimensional tolerances can dictate the processing. In any case, in the sequence of converting the design into reality, such decisions must be made by someone. In most instances they can be made most effectively at the design stage by the designer if he has a reasonably adequate knowledge concerning materials and manufacturing processes. Otherwise, decisions may be made that will detract from the effectiveness of the product, or the product may be needlessly costly. It is thus apparent that design engineers are a vital factor in the manufacturing process, and it is indeed a blessing to the company if they can design for producibility—that is, for efficient production.

Manufacturing engineers select and coordinate specific processes and equipment to be used, or supervise and manage their use. Some design specific tooling that is used so that standard machines can be utilized in producing specific products. These engineers must have a broad knowledge of machine and process capabilities and of materials, so that desired operations can be done effectively and efficiently without overloading or damaging machines and without adversely affecting the materials being processed. These manufacturing engineers also play an important role in manufacturing.

A relatively small group of engineers design the machines and equipment used in manufacturing. They obviously are design engineers and, relative to their products, have the same concerns of the interrelationship of design, materials, and manufacturing processes. However, they have an even greater concern regarding the properties of the materials that their machines are going to process and the interaction of the materials and the machines.

Still another group of engineers—the materials engineers—devote their major efforts to developing new and better materials[1]. They, too, must be concerned with how these materials can be processed and with the effects the processing will have on the properties of the materials.

Although their roles may be quite different, it is apparent that a large proportion of engineers must concern themselves with the interrelationship between materials and manufacturing processes.

Low-cost manufacture does not just happen. There is a close and interdependent relationship between the design of a product, selection of materials, selection of processes and equipment, and tooling selection and design. Each of these steps must be carefully considered, planned, and coordinated before manufacturing starts. This lead time, particularly for complicated products, may take months, even years, and the expenditure of large amount of money may be involved. Typically, the lead time for a completely new model of an automobile is about 2 years. For a modern aircraft it may be 4 years.

With the advent of computers and machines that can be controlled by either tapes made by computers or by the computers themselves, we are entering a new era of production planning. The integration of the design function and the manufacturing function through the computer is called CAD/CAM (computer aided design/computer aided manufacturing). The design is used to determine the manufacturing process planning and the programming information for the manufacturing processes themselves. Detailed drawings can also be made from the central data base used for the design and manufacture, and programs can be generated to make the parts as needed[2]. In addition, extensive computer aided testing and inspection (CATI) of the manufactured parts is taking place. There is no doubt that this trend will continue at ever-accelerating rates as computers become cheaper and smarter.

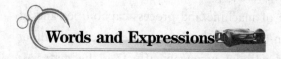 **Words and Expressions**

dictate   [dɪkˈteɪt]   v.   指示,规定

procedure	[prə'siːdʒə]	n.	(程序,工序)生产过程
sequence	['siːkwəns]	n.	顺序,程序,工序
detract	[dɪtrækt]	v.	转移,贬低
coordinate	[kəʊ'ɔːdɪnət]	v.	以使协调,整理
tooling	['tuːlɪŋ]	n.	工具,刀具,工艺装备
interdependent	[ˌɪntədɪ'pendənt]	adj.	相互依赖的,相互影响的
interrelationship	[ˌɪntərɪ'leɪʃɪnʃɪp]	n.	相互关系
advent	['ædvənt]	n.	到来,出现,来临
integration	[ˌɪntɪ'greɪʃn]	n.	集成,综合,集体化
modify	['mɒdɪfaɪ]	v.	改变,改善,调整
needlessly	['niːdlɪsli]	adv.	不必要地,无用地
complicated	['kɒmplɪkeɪtɪd]	adj.	错综复杂的,麻烦的
expenditure	[ɪk'spendɪtʃə]	n.	消费,支出
inspection	[ɪn'spekʃn]	n.	检验,观察,验收
ever-accelerating		adj.	不断加速的
lead time			产品设计至实际投产所需的时间

NOTES

[1] Still another group of engineers—the materials engineers—devote their major efforts to developing new and better materials.

还有另外一些工程师,即材料工程师,他们致力于开发更好的新型材料。(Devote to 是"致力于……"的意思)

[2] Detailed drawings can also be made from the central data base used for the design and manufacture, and programs can be generated to make the parts as needed.

设计与制造的中心数据库可以绘制详细的零件图,并生成这些零件所需的加工程序。

第1单元 工程师在生产制造中的作用

许多工程师的职责是设计产品,并通过对材料的加工制造出产品。在选择产品的材料和制造方法时,设计工程师起着关键的作用。设计工程师应该比其他人更清楚他的设计目的。他应该对工作载荷和使用要求做出正确的假设,了解产品的使用环境,并确定产品的外观。因此,他必须选择和确定产品使用的材料。通常,为了充分利用材料并加工出理想的产品,他还应该熟悉那些生产制造中必须采用的工艺。许多情况下,材料的选择就决定了加工工艺的方式。同时,采用了某种加工工艺,设计也必须做相应的修改,以确保所采用的工艺能够提高效率、降低成本。尺寸公差也能影响加工方法。任何情况下,要将设计转变为现实,就必须做出这些决断。多数情况下,如果设计人员充分掌握了材料和加工方法的有关知识,他就会做出效率最高的设计。否则,就会使产品的效能降低、价格陡增。显然,设计工程师是制造过程中的重要人物。如果他们的设计面向生产,那确实是公司的幸事,因为这可以提高生产的效率。

制造工程师选择并协调将要使用的加工方法和加工设备,或者监督和管理这些加工方法和加工设备的使用。某些工程师设计专用工艺装备,使通用机床也能生产专用产品。这些工程师应该在机器、加工能力及材料方面具有广博的知识,使生产高效进行,既不因过载损坏机器,也不会对加工材料产生不良影响。制造工程师在制造业中也起到重要的作用。

还有一小部分工程师专门设计机床和设备,他们显然属于设计工程师。相对于他们的产品,他们同样关心设计、材料和制造方法之间的相互关系。但是,他们更关心他们的机器将要加工的材料的性能及机器与材料之间的相互作用。

还有一些工程师,即材料工程师,他们致力于开发更好的新型材料,他们同样也关心材料的加工方法及其对材料性能的影响。

尽管各类工程师的作用千差万别,但显然他们多数都要考虑材料与制造工艺之间的相互关系。

降低制造成本可不是件容易的事情。产品设计、材料选择、加工工艺和设备的选择、工具和设计的选择等之间存在非常密切的相互依赖关系。每一步在生产开始前都必须仔细考虑、精心计划、互相协调。从产品设计到实际投产,特别是复杂产品,可能需要数月甚至数年的时间,要花费很多钱。例如,对于一种全新的汽车,从产品设计到实际投产需要大约2年的时间,而一种现代化飞机则可能需要4年。

随着计算机的出现以及由穿孔纸带控制或由计算机本身控制的机器的出现,我们正进入一个生产计划的新时代。采用计算机把产品的设计与制造结合起来,称为CAD/CAM(计算机辅助设计/计算机辅助制造)。用这种设计方法可以确定加工工艺计划以及与加工工艺本身有关的编程信息。设计与制造的中心数据库可以绘制详细的零件图,并生成这些零件所需的加工程序。此外,零件加工还使用了大量的计算机辅助实验与计算机辅助检测。毫无疑问,随着计算机价格的降低和性能的提高,这种趋势将会持续飞速发展。

READING MATERIAL

The Machine Designer's Responsibility

A new machine is born because there is a real or imagined need for it. It evolves from someone's conception of a device with which to accomplish a particular purpose. From the conception follows a study of the arrangement of the parts, the location and length of links (which may include a kinematic study of the linkage), the places for gears, bolts, springs, cams, and other elements of machines. With all ideas subject to change and improvement, several solutions may be and usually are found, the seemingly best one being chosen.

The actual practice of designing is applying a combination of scientific principles and a knowing judgment based on experience. It is seldom that a design problem has only one right answer, that a situation that is often annoying to the beginner in machine design.

Engineering practice usually requires compromises. Competition may require a reluctant decision contrary to one's best engineering judgment; production difficulties may force a change of design; etc.

A good designer needs many attributes, for example:

(1) A good background in strength of materials, so that the stress analyses are sound. The parts of the machine should have adequate strength and rigidity, or other characteristics as needed.

(2) A good acquaintance with the properties of materials used in machines.

(3) A familiarity with the major characteristics and economics of various manufacturing processes, because the parts that make up the machine must be manufactured at a competitive cost. It happens that a design that is economic for one manufacturing plant may not be so for another. For example, a plant with a well-developed welding department but no foundry might find that welding is the most economic fabricating method in a particular situation; whereas another plant faced with the same problem might decide upon casting because they have a foundry (and may or may not have a welding department).

(4) A specialized knowledge in various circumstances, such as the properties of materials in corrosive atmospheres, at very low (cryogenic) temperatures, or at relatively high temperatures.

(5) A preparation for deciding wisely: ① when to use manufacturers' catalogs, buying stock or relatively available items, and when custom design is necessary; ② when empirical design is justified; ③ when the design should be tested in service tests before manufacture starts; ④ when special measures should be taken to control vibration and sound (and others).

(6) Some aesthetic sense, because the product must have "customer appeal" if it is to sell.

(7) A knowledge of economics and comparative costs, because the best reason for the existence of engineers is that they save money for those who employ them. Anything that increases the cost should be justified by, for instance, an improvement in performance, the addition of an attractive feature, or greater

durability.

(8) Inventiveness and the creative instinct, most important of all for maximum effectiveness. Creativeness may arise because an energetic mind is dissatisfied with something as it is and this mind is willing to act.

Naturally, there are many other important considerations and a host of details. Will the machine be safe to operate? Is the operator protected from his own mistakes and carelessness? Is vibration likely to cause trouble? Will the machine be too noisy? Is the assembly of the parts relatively simple? Will the machine be easy to service and repair?

Of course, no one engineer is likely to have enough expert knowledge concerning the above attributes to make optimum decisions on every question[1]. The larger organizations will have specialists to perform certain functions, and smaller ones can employ consultants. Nevertheless, the more any one engineer knows about all phrases of design, the better. Design is an exacting profession, but highly fascinating when practiced against a broad background of knowledge.

Words and Expressions

arrangement	[əˈreɪndʒmənt]	n.	配置,布局,构造
linkage	[ˈlɪŋkɪdʒ]	n.	连杆,连杆机构
seemingly	[ˈsiːmɪŋli]	adv.	表面上,外观上,看上去
kinematic	[ˌkaɪnɪˈmætɪk]	adj.	运动学的
annoy	[əˈnɔɪ]	vt.	使……烦恼
compromise	[ˈkɒmprəmaɪz]	n.	妥协
reluctant	[rɪˈlʌktənt]	adj.	不愿意的,勉强的,难得到的
attribute	[ˈætrɪbjuːt]	n.	属性,特性,特征
acquaintance	[əˈkweɪntəns]	n.	熟悉,了解
foundry	[ˈfaʊndri]	n.	铸造
fabricate	[ˈfæbrɪkeɪt]	vt.	制作,制造
atmosphere	[ˈætməsfɪə]	n.	大气层,空气
cryogenic	[ˌkraɪəˈdʒenɪk]	adj.	冷冻的,低温的
stock	[stɒk]	n.	原料,材料
empirical	[emˈpɪrɪkl]	adj.	经验的,实验的
aesthetic	[iːsˈθetɪk]	adj.	审美的,美学的,美术的
appeal	[əˈpiːl]	n.	吸引力,感染力
comparative	[kəmˈpærətɪv]	adj.	比较的,相当的
justify	[ˈdʒʌstɪfaɪ]	v.	证明,认为……有理由

inventiveness	[ɪnˈventɪvnəs]	n.	发明创造能力,创造性
instinct	[ˈɪnstɪŋkt]	n.	本性,本能,直觉
energetic	[enəˈdʒetɪk]	adj.	有活力的,精力旺盛的
host	[həʊst]	n.	许多,大量
fascinating	[ˈfæsɪneɪtɪŋ]	adj.	引人入胜的,极有趣的

NOTE

...no one engineer is likely to have enough expert knowledge concerning the above attributes to make optimum decisions on every question.

没有一个工程师能够具有上述那些基本素质所应有的充分的专业知识,对每一个问题作出最适宜的解答。(be likely to... 是"大概","可能"的意思)

阅读材料

机械设计者的责任

新机械产生于实际的或设想的需要。它来自设计者为了达到某个具体目的而准备设计的机械设施的概念。根据这个概念,还有零件布置的研究,连接件的长度和位置的研究(可能还要包括连杆的运动学研究),齿轮、螺栓、弹簧、凸轮及其他机械零件的位置的研究。通过修正和改进我们的想法,我们可能并通常可以找到几个解决方案,从中选择一个看起来最好的。

实际的设计活动是各种科学原理的综合应用,是基于已有经验的正确判断。设计问题很少只有一种正确的答案,这一点往往困扰着机械设计新手。

设计经常要向工程实际妥协,为满足竞争需要而勉强做出的决定可能与人们最好的工程判断完全相反,生产制造的困难也会使我们不得不改变设计,等等。

优秀的设计者应该具备许多基本素质,包括:

1. 掌握材料力学方面的知识,以便能够进行充分的应力分析,使机械零件具有足够的强度、刚度及其他所需的特性。

2. 熟知机械所用材料的特性。

3. 熟悉各种制造工艺的特点及成本,因为组装机器的零件的制造成本必须具有竞争性。在一个工厂碰巧被证明为经济实用的设计方案,可能并不适用于另一个工厂。例如,有的工厂焊接部门很先进但却没有铸造部门。在这种特殊条件下,他们可能会认为焊接是最经济的制造手段,而对于那些有铸造部门(可能有也可能没有焊接部门)的工厂,面临同样的问题却可能会选择铸造。

4. 有关各种工作环境的专业知识。例如,在腐蚀气体、低温或相对高温下材料的特性。

5. 精明决断的必要准备。要了解：① 什么时间按照厂商目录购买原料及相关物品；什么时候必须按用户的要求确定设计。② 什么时候根据经验设计。③ 什么时候进行投产前的实验。④ 什么时候要采取特殊措施控制振动和噪声。

6. 一定的审美能力，设计的产品销售时能够吸引顾客。

7. 要了解各种经济学和成本比较的知识。对雇主来说，雇佣工程师最主要的原因是他们能够为雇主节省费用。成本增加必须有充分的理由，例如，改进性能，增加吸引力，或使产品更经久耐用。

8. 与生俱来的创意与创造灵感，这是提高效率的最重要素质。创造性来源于不满足现状，并使思想付诸实施。

当然，还有许多其他重要的因素和细节。例如，机器操作是否安全？操作者粗心大意是否不会受到伤害？振动可能引起故障吗？机器的噪声很大吗？零件的装配相对简便吗？机器使用和维修容易吗？

当然，没有一个工程师能够具有上述那些基本素质所应有的充分的专业知识，对每一个问题作出最适宜的解答。大公司有各个行业的专家，小公司可以聘请顾问。但工程师对设计各方面的了解，总是越多越好。设计是一个要求严格的职业，但只要具有丰富的知识背景，设计实践工作还是十分有趣的。

Unit 2

Selection of Materials

A material is generally used because it offers the required properties at reasonable cost. Appearance is also an important factor. Perhaps the most common classification that is encountered in materials selection is whether the material is metallic or nonmetallic. The common metallic materials are such metals as iron, copper, aluminum, magnesium, nickel, titanium, lead, tin, and zinc and the alloys of these metals, such as steel, brass, and bronze. The metallic material is further classified as ferrous (iron and its alloys) and non-ferrous (all other metallic materials) metal. They possess the metallic properties of luster, thermal conductivity, and electrical conductivity; and they are relatively ductile; and some have good magnetic properties. The common nonmetals are wood, brick, concrete, glass, rubber, and plastics. Their properties vary widely, but they generally tend to be less ductile, weaker, and less dense than the metals, and they have no electrical conductivity and poor thermal conductivity.

Although it is likely that metals always will be the more important of the two groups, the relative importance of the nonmetallic group is increasing rapidly, and since new nonmetals are being created almost continuously, this trend is certain to continue. In many cases the selection between a metal and nonmetal is determined by a consideration of required properties. Where the required properties are available in both, total cost becomes the determined factor[1].

One material can often be distinguished from another by means of physical properties, such as color, density, specific heat, coefficient of thermal expansion, thermal and electrical conductivity, magnetic properties, and melting point. Some of these, for example, thermal conductivity, electrical conductivity, and density, may be of prime importance in selecting material for certain specific uses. However, those properties that describe how a material reacts to mechanical usages are often more important to the engineers in selecting materials in connection with design. These mechanical properties relate to how the material will react to the various loading service.

Mechanical properties are the characteristic response of materials to applied

forces. These properties fall into five broad categories: strength, hardness, elasticity, ductility, and toughness.

(1) Strength is the ability of a material resist applied forces. Elevator cables and buildings beams all must have this property.

(2) Hardness is the ability of a material resist penetration and abrasion. Cutting tools must resist abrasion, or wear. Metal rolls for steel mills must resist penetration.

(3) Elasticity is the ability to spring back to original shape. All springs should have this quality.

(4) Ductility is the ability to undergo permanent changes of shape without rupturing. Stamped and formed products must have this property.

(5) Toughness is the ability to absorb mechanically applied energy. Strength and ductility determine a material's toughness. Toughness is needed in railroad cars, automobile axles, hammers, and similar products.

The main advantage of metals is their strength and toughness. Concrete may be cheaper and is often used in building, but even concrete depends on its core of steel for strength. Not all metals are strong, however. Copper and aluminum, for example, are both fairy weak, but if they are mixed together, results in an alloy called as aluminum-bronze alloy, which is much stronger than either pure copper or pure aluminum. Alloying is an important method of obtaining whatever special properties required: strength, toughness, resistance to wear, magnetic properties, high electrical resistance or corrosion resistance[2].

Plastics have specific properties, which may make them preferable to traditional materials for certain uses. In comparison with metals, for example, plastics have both advantages and disadvantages. Metals tend to be corroded by inorganic acids, such as sulphuric acid and hydrochloric acid. Plastics tend to be resistant to these acids, but can be dissolved or deformed by solvents, such as carbon tetrachloride, which have the same carbon bases as the plastics[3]. Metals are more rigid than most plastics, while plastics are very light, with a specific gravity normally between 0.9 and 1.8. More plastics do not readily conduct heat or electricity. Plastics soften slowly and can easily be shaped while they are soft.

It is their plasticity at certain temperatures, which give plastics their main advantages over many other materials. It permits the large-scale production of molded articles, such as containers, at an economic unit cost, while other materials require laborious and often costly processes involving casting, shaping, machining, assembly and decoration. Plastics are lighter and more corrosion-resistant, but they are not usually as strong. Another problem with plastics is what to do with them after use. Metal objects can often be broken down and the metals are recycled; plastics can only be dumped or burned.

Words and Expressions

metallic	[mɪˈtælɪk]	adj.	金属的
iron	[ˈaɪən]	n.	铁
copper	[ˈkɒpə]	n.	铜,紫铜
aluminum	[əˈljuːmɪnəm]	n.	铝
magnesium	[mægˈniːzɪəm]	n.	镁
nickel	[ˈnɪkl]	n.	镍
titanium	[tɪˈteɪnɪəm]	n.	钛
lead	[liːd]	n.	铅
tin	[tɪn]	n.	锡
zinc	[zɪŋk]	n.	锌
brass	[brɑːs]	n.	黄铜
bronze	[brɒnz]	n.	青铜
ferrous metal			黑色金属
non-ferrous metal			有色金属
brick	[brɪk]	n.	砖头
concrete	[ˈkɒŋkriːt]	n.	混凝土
luster	[ˈlʌstə]	n.	光泽
thermal conductivity			导热性
density	[ˈdensəti]	n.	密度
specific heat			比热
coefficient of thermal expansion			热膨胀系数
ductility	[dʌkˈtɪlɪti]	n.	延展性
toughness	[ˈtʌfnəs]	n.	硬度
elevator	[ˈelɪveɪtə]	n.	电梯
abrasion	[əˈbreɪʒən]	n.	磨损
rupture	[ˈrʌptʃə]	v.	破裂
inorganic	[ˌɪnɔːˈgænɪk]	adj.	无机的
sulphuric acid			硫酸
hydrochloric acid			盐酸
solvent	[ˈsɒlvənt]	n.	溶剂
carbon tetrachloride			四氯化碳

NOTES

[1] Where the required properties are available in both, total cost becomes the determined factor.

在两种材料都能提供所需性能时,总成本就成了决定性因素。(句中 where 引导状语从句,这一段讨论的是金属和非金属两组材料,所以句中的 both 是指这两组中的双方)

[2] Alloying is an important method of obtaining whatever special properties... corrosion resistance.

合金是获得各种所需特殊性能(强度、韧性、耐磨性、磁性、高电阻或耐腐蚀性)的重要方法。(句中 whatever 是"不管什么""诸如此类"的意思。冒号引起罗列各种相关的性能,相当于 such as)

[3] Plastics tend to be..., which have the same carbon bases as the plastics.

塑料能抵抗这些酸,但会被诸如同样有酸基的四氯化碳这样的溶剂溶解或致变形。(注意到句中采用的是非限定性定语从句,其中的动词是复数形式。这个定语修饰的是 solvents 而不是 carbon tetrachloride)

第 2 单元　材料的选用

材料通常因为它能以合理的成本提供需要的性能而被选用。外观也是一个重要的因素。材料选择中碰到的最普通的分类或许在于材料是金属还是非金属。常用的金属材料有铁、铜、铝、锰、钛、铅、锡、锌和它们的合金,如钢、黄铜、青铜。金属材料被进一步分为黑色金属(铁及合金)和有色金属(所有其他金属材料)。黑色金属具有金属的导热性、导电性等特性,外表光泽相对而言延展性好,有些还具有良好的磁性。常用的非金属有木材、砖、混凝土、玻璃、橡胶和塑料。它们的性能变化很大,但通常趋于延展性差、强度低、比金属密度小、无导电性且导热性差。

尽管金属可能总是在这两类材料中更为重要,但非金属的相对重要性正在迅速攀升,并且由于不断有新的非金属产生,这种趋势必定还将继续。在许多情况下,在金属和非金属之间的选择根据对所需性能的考虑来决定。在两种材料都能提供所需性能时,总成本就成了决定性因素。

一种金属常常通过物理性能来区别于另一种金属,例如颜色、密度、比热、热膨胀系数、导热性和导电性、磁性和熔点。在为某一特定用途选用材料时,某些性能,如导热性、导电性和密度,可能是首要的。材料对各种机械用途所表现的性能的反应,对机械工程师在设计选材时往往更为重要。这些性能关系到材料对各种负载工况的反应。

力学性能是材料对所施外力的响应特征。这些性能分为五大类:强度、硬度、弹性、延展性和韧性。

1. 强度是材料抵抗外力的能力。电梯吊缆和建筑梁都有这样的性能。

2. 硬度是材料抵抗渗透和耐磨损的能力。刀具必须要耐磨损,钢铁碾磨机的金属滚轮必须能抵抗渗透。

3. 弹性是弹回到原始形状的能力。所有的弹簧必须有这种性能。

4. 延展性是承受永久变形而不破裂的能力。冲压成型产品必须有这种性能。

5. 韧性是吸收机械能量的能力。强度和延展性决定材料的韧性。铁路车辆、汽车车桥、锻锤及类似产品需要这种性能。

金属的主要优点在于它们的强度和韧性。混凝土也许因价廉而经常用于建筑,但需依赖钢筋提高强度。并非所有的金属都有高强度,例如,紫铜和铝的强度都低,但如果它们混合在一起,生成一种称为铝青铜的合金,就比纯铜或纯铝的强度高得多。合金是获得各种所需特殊性能(强度、韧性、耐磨性、磁性、高电阻或耐腐蚀性)的重要方法。

塑料有特殊的性能,这使它们对某些特定用途比传统材料更可取。例如,与金属相比,塑料有优点也有缺点。金属易于被诸如硫酸、盐酸等无机酸锈蚀。塑料能抵抗这些酸但会被诸如同样有酸基的四氯化碳这样的溶剂溶解或致变形。金属比大多数塑料刚性更好,而塑料非常轻,相对密度在0.9~1.8之间。大多数塑料不能导热或导电,软化缓慢,但极易在软化状态成型。

正是在特定温度下的塑性,赋予了塑料超越其他材料的主要优点,如容器之类的模压产品在大规模生产时使用塑料可降低单价,使用其他材料往往耗时费力,其昂贵的加工过程包括铸造、成型、机加工、装配和装饰。塑料更轻、更耐腐蚀,但强度低。另一个问题是塑料用过之后怎么处理。金属通常可以解体回收;塑料只能丢弃或焚烧。

READING MATERIAL

Heat Treatment

Heat treatment is thermal cycling involving one or more reheating and cooling operations after forging for the purpose of obtaining desired microstructures and mechanical properties in a forging.

Few forgings of the types are produced without some form of heat treatment. Untreated forgings are usually relatively low-carbon steel parts for noncritical applications or are parts intended for further hot mechanical work and subsequent heat treatment. The chemical composition of the steel，the size and shape of the product，and the properties desired are important factors in determining which of the following production cycles to use.

The object of heat treating metals is to impart certain desired physical properties to the metal or to eliminate undesirable structural conditions which may occur in the processing or fabrication of the material. In the application of any heat

treatment it is desirable that the "previous history" or structural condition of the material be known so that a method of treatment can be prescribed to produce the desired result. In the absence of information as to the previous processing, a microscopic study of the structure is desirable to determine the correct procedure to be followed.

The commercial heat treatments in common use involve the heating of the material to certain predetermined temperatures, "soaking" or holding at the temperature, and cooling at a prescribed rate in air, liquids, or retarding media[1].

Spheroidizing is heating of iron-based alloys at a temperature slightly below the critical temperature range followed by relatively slow cooling, usually in air. Small objects of high carbon steel are more rapidly spheroidized by prolonged heating to temperatures alternately within and slightly below the critical temperature range[2]. The purpose of this heat treatment is to produce a globular condition of the carbide.

Normalizing is heating iron-base alloys to temperatures approximately 50℃ above the critical temperature range followed by cooling in air to below the range. The purpose is to put the metal structure in a normal condition by removing all internal strains and stresses given to the metal during some processing operation.

Hardening is a process to increase its hardness and tensile strength, to reduce its ductility, and to obtain a fine grain structure. The procedure includes heating the metal above its critical point of temperature, followed by rapid cooling. As steel is heated, a physical and chemical change takes place between the iron and carbon. The critical point, or critical temperature, is the point at which the steel has the most desirable characteristics. When steel reaches this temperature, somewhere between 1 400 and 1 600°F, the change is ideal to make for a hard, strong material if it is cooled quickly. If the metal cools slowly, it changes back to its original state. By plunging the hot metal into water, oil, or brine (quenching), the desirable characteristics are retained. The metal is very hard and strong and less ductile than before.

Steel that has been hardened by rapid quenching is brittle and not suitable for most uses. By tempering or "drawing", the hardness and brittleness may be reduced to the desired point for service conditions. As these properties are reduced, there is also a decrease in tensile strength and an increase in the ductility and toughness of the steel. The operation consists of the reheating of quench-hardened steel to some temperature below the critical range, followed by any rate of cooling. Although this process softens steel, it differs considerably from annealing in that the process lends itself to close control of the physical properties and in most cases does not soften the steel to the extent that annealing would. The final structure obtained from tempering fully hardened steel is called tempered martensite.

Tempering is possible because of the instability of the martensite, the principal constituent of hardened steel. Low temperature draws, from 300 to 400°F(150 to 205℃), do not cause much decrease in hardness and are used principally to relieve internal strains. As the tempering temperatures are increased, the breakdown of the martensite takes place at a faster rate, and at about 600°F(315℃) the change to a structure called tempered martensite is very rapid. The tempering operation may be described as one of precipitation and agglomeration, or coalescence of cementite. A substantial precipitation of cementite is at 600°F(315℃), which produces a decrease in hardness. Increasing the temperature causes coalescence of the carbides, with continued decrease in hardness.

The primary purpose of annealing is to soften hard steel so that it may be machined or cold-worked. This is usually accomplished by heating the steel to slightly above the critical temperature to form austenite, holding it there until the temperature of the piece is uniform throughout, and then cooling at a slowly controlled rate so that temperature of the surface and that of the center of the piece are approximately the same. This process is known as full annealing because it wipes out all trace of previous structure, refines the crystalline structure, and softens the metal. Annealing also relieves internal stresses previously set up in the metal.

When hardened steel is reheated to above the critical range, the constituents are changed back into austenite, and slow cooling then provides ample time for complete transformation of the austenite into the softer constituents. For the hypoeutectoid steels, these constituents are pearlite and ferrite.

The temperature to which given steel should be heated in annealing depends on its composition, and for carbon steels it can be obtained readily from the iron-carbide equilibrium diagram. The heating rate should be consistent with the size of sections so that the center part is brought up to temperature as uniform as possible. When the annealing temperature has been reached, the steel should be held there until it is uniform throughout. For maximum softness and ductility, the cooling rate should be very slow, such as allowing the parts to cool down with the furnace. The higher the carbon content, the slower this rate must be.

Words and Expressions

tempering	['temperɪŋ]	n.	回火
microstructure	['maɪkrəˌstrʌktʃə]	n.	显微组织
soaking	[səʊkɪŋ]	n.	均热,保温
prescribe	[prɪ'skraɪb]	v.	规定,指示

prolonged	[prə'lɒŋd]	*adj.*	长时间的,持续很久的
microscopic	['maɪkrə'skɒpɪk]	*adj.*	显微镜的
spheroidizing	['sfɪərɔɪdaɪzɪŋ]	*n.*	球化(处理)
globular	['glɒbjʊlə]	*adj.*	球形(状)的
normalizing	['nɔ:məlaɪzɪŋ]	*n.*	正火,(正)常化
carbide	['kɑ:baɪd]	*n.*	碳化物,硬质合金
plunging	['plʌndʒɪŋ]	*adj.*	跳进的,突进的
brine	[braɪn]	*n.*	盐水
annealing	[ə'ni:lɪŋ]	*n.*	退火
quenching	[kwəntʃɪŋ]	*n.*	淬火,骤冷
hardening	['hɑ:dənɪŋ]	*n.*	淬火
removal	[rɪ'mu:vəl]	*n.*	除去,放出
martensite	['mɑ:tənzaɪt]	*n.*	马氏体
precipitation	[prɪˌsɪpɪ'teɪʃn]	*n.*	沉淀
agglomeration	[əˌglɒmə'reɪʃn]	*n.*	结块,凝聚,块
cementite	[sɪ'mentaɪt]	*n.*	渗碳体,碳化铁
austenite	['ɔ:stəˌnaɪt]	*n.*	奥氏体
hypoeutectoid	[ˌhaɪpəʊju'tektɔɪd]	*adj.*	亚共析的
pearlite	['pɜ:laɪt]	*n.*	珠光体
ferrite	['feraɪt]	*n.*	铁素体
in that			由于,因为,既然

NOTES

[1] The commercial heat treatments in common use involve the heating of the material to certain predetermined temperatures,"soaking" or holding at the temperature,and cooling at a prescribed rate in air,liquids,or retarding media.

经常使用的商业热处理是指把材料加热到某一预定的温度,在此温度下进行均热,即进行保温,并按规定速度在空气、液体或在保温介质中冷却。(句中的 the heating…"soaking"… and cooling 是 involve 的宾语;or holding 是 soaking 的同位语,or 译为"即")

[2] Small objects of high carbon steel… the critical temperature range.

在临界温度和略低于临界温度的范围内长时间重复加热,小型高碳钢零件可以更迅速地球化。

热处理

热处理是锻后一次或多次重新加热和冷却操作的热循环过程,使锻件获得所需的显微组织和力学性能。

几乎所有锻件都需要进行某种形式的热处理。没有经过热处理的锻件,要么是应用场合不太重要的碳含量相对较低的零件,要么是那些需要进一步热加工后再热处理的锻件。钢的化学成分、产品的规格和形状以及所希望的性能是选择接下来所使用的生产工艺的重要因素。

金属热处理的目的是使金属获得所需的物理性能,或者是消除那些在材料生产和加工中可能出现的组织结构缺陷。应用任何热处理方法都应该知道"以前的历史"或材料的组织状态,这样才能够确定一种具体的处理方法并达到预期的结果。如果缺乏前面加工的有关信息,必须对组织结构进行显微研究,以便确定要采用的正确的步骤。

经常使用的商业热处理是指把材料加热到某一预定的温度,在此温度下进行均热,即进行保温,并按规定速度在空气、液体或保温介质中冷却。

球化处理是指在略低于临界温度的条件下,对铁基合金持续长时间加热,紧接着以较慢的速度在空气中冷却。在临界温度和略低于临界温度的范围内长时间重复加热,小型高碳钢零件可以更迅速地球化。这种热处理的目的是生成球状碳化物组织。

正火是指把铁基合金加热到临界温度50℃以上,紧接着在低于临界温度下自然冷却。其目的是消除金属的组织结构中因某种加工操作产生的全部内应力,以便使金属恢复到正常状态。

淬火工艺可以增加材料的硬度、抗拉强度,减少其延展性并获得细微的晶格,方法是把金属加热到材料的临界温度之上后迅速冷却。加热时,铁和碳之间发生物理化学反应。而在临界点或者临界温度,钢可以获得最理想的特性。当钢被加热到1400°F～1600°F(760℃—871℃)后,如果迅速冷却,将有利于生成一种坚硬、强韧的材料。如果慢慢冷却,金属会回变成原来的状态。把热金属投入水、油或者盐水(淬火液)中,金属理想的特性被保持——非常坚硬、强韧,但延展性比原来有所降低。

迅速淬火的钢很脆,应用十分有限,回火可以降低硬度和脆性,使材料满足工作条件。随着这些特性的降低,钢的抗拉强度同时降低,但延展性和韧性有所提高。回火操作是指把淬过火的钢重新加热到临界温度以下,然后以任意的速度冷却。虽然这种工艺使钢变软了,但与退火不同的是,这种工艺本身可以精密控制材料的物理特性,使钢在大多数情况下要硬一些。充分淬火的钢经回火所得的最终结构称为回火马氏体。

淬过火的钢的主要成分是马氏体,由于不太稳定,因而可以进行回火处理。低温回火的温度为300°F～400°F(150℃～205℃),硬度降低不多,主要用于释放内应力。随着回火温度的增加,马氏体的分解速度越来越快,在大约600°F(315℃)时迅速变成回火马氏体。回火操作可以描述为渗碳体的沉淀、积聚或结合。渗碳体在600°F(315℃)

时充分积聚,硬度降低。增加温度碳素体继续积聚,硬度继续降低。

退火的主要目的是使硬度较高的钢变软,以利于机加工或冷加工。通常的操作是把钢稍微加热到临界温度以上形成奥氏体并保温,当整个零件的温度完全均衡后再慢慢冷却,以保证零件表面与中心的温度大致相同。这种工艺称之为完全退火,因为它消除了所有前面工序遗留的结构缺陷,细化了晶格,使金属变软。退火还可以释放金属中残留的内应力。

当淬过火的钢被重新加热到临界温度之上时,其成分会变成奥氏体,此时缓慢冷却可以提供充分的时间把奥氏体完全转换成为较软的组织。对于低碳钢(亚共析钢),这些组织就是珠光体或铁素体。

退火时的加热温度依赖于钢的成分,碳素钢的加热温度从铁碳平衡图中很容易获得。加热的速度应该与零件截面的大小一致,以便使零件中心部分的温度尽可能均衡。当加热到退火温度时,应该保温,使零件彻底热透。为了尽可能使钢变软并获得最好的延展性,冷却速度应该十分缓慢,例如可以随加热炉一起冷却。碳的含量越高,冷却速度必须越慢。

Unit 3

Lathes

扫码可见本项目
☞ 相关参考资料

A machine tool performs three major functions:

（1）it rigidly supports the workpiece or its holder and the cutting tool;

（2）it provides relative motion between the workpiece and the cutting tool;

（3）it provides a range of feeds and speeds.

Machines used to remove metal in the form of chips are classified as follows:

（1）Machines using basically the single-point cutting tools include: engine lathes, turret lathes, tracing and duplicating lathes, single-spindle automatic lathes, multi-spindle automatic lathes, shapers and planers, boring machines.

（2）Machines using multipoint cutting tools include: drilling machines, milling machines, broaching machines, sawing machines, gear-cutting machines.

（3）Machines using random-point cutting tools(abrasive) include: cylindrical grinders, centreless grinders, and surface grinders.

Special metal removal methods include: chemical milling, electrical discharge machining, ultrasonic machining.

The lathe removes material by rotating the workpiece against a cutter to produce external or internal cylindrical or conical surfaces. It is commonly used for the production of surfaces by facing, in which the work piece is rotated while the cutting tool is moved perpendicularly to the axis of rotation.

The engine lathe, shown in Figure 3 - 1, is the basic turning machine from which other turning machines have been developed. The driving motor is located in the base and drives the spindle through a combination of belts and gears. The spindle is a sturdy hollow shaft, mounted between heavy-duty bearings, with the forward end used for mounting a drive plate to impart positive motion to the workpiece. The drive plate may be fastened to the spindle by threads, by a cam lock mechanism or by a threaded collar and key.

The lathe bed is cast iron and provides accurately ground sliding surfaces(way) on which the carriage rides[1]. The lathe carriage is an H-shaped casting on which the cutting tool is mounted in a tool holder. The apron hangs from the front of the carriage and contains the driving gears that move the tool and carriage along or across the way to provide the desired tool motion.

A compound rest, located above the carriage, provides for rotation of the tool holder through any desired angle. A hand wheel and feed screw are provided on the compound rest for linear motions of the tool. The cross feed is provided with another hand wheel and feed screw for moving the compound rest perpendicular to the lathe way. A gear train in the apron provides power feed for the carriage both along and across the way. The feed box contains gears to impart motion to the carriage and control the rate at which the tool moves relative to the workpiece. Since the transmission in the feed box gearing is driven by the spindle gears, the feeds are directly related to spindle speed. The feed box gearing is also used in thread cutting and provides from 4 to 224 threads per in.

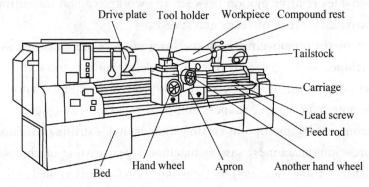

Figure 3 - 1   Engine lathe

The turret lathe is basically an engine lathe with certain additional features to provide for semiautomatic operation and to reduce the opportunity for human error. The carriage of the turret lathe is provided with T-slots for mounting a tool-holding device on both sides of the lathe ways with tools properly set for cutting when rotated into position. The carriage is also equipped with automatic stops that control the tool travel and provide good reproduction of cuts. The tailstock of the turret lathe is of hexagonal design, in which six tools can be mounted. Although a large amount of time is consumed in setting up the tools and stops for operation, the turret lathe, once set, can continue to duplicate operations with a minimum of operator skill until the tools become dulled and need replacing. Thus, the turret lathe is economically feasible only for production work, where the amount of time necessary to prepare the machine for operation is justifiable in terms of the number of part to be made.

The multi-spindle automatic lathe is provided with four, five, six, or eight spindles, with one workpiece mounted in each spindle. The spindles index around a central shaft, with the main tool slide accessible to all spindles. Each spindle position is provided with a side tool-slide operated independently. Since all of the slides are operated by cams, the preparation of this machine may take several days, and a production run of at least 5 000 parts is needed to justify its use. The principal advantage of this machine is that all tools work simultaneously, and one operator can handle several machines. For relatively simple parts, multi-spindle automatic lathes can turn out finished products at the rate of 1 every 5 sec.

Words and Expressions

lathe	[leɪð]	n.	车床
chip	[tʃɪp]	n.	切屑
shaper	[ˈʃeɪpə]	n.	牛头刨床
planer	[ˈpleɪnə(r)]	n.	龙门刨床
abrasive	[əˈbreɪsɪv]	adj.	摩擦的,磨损的
sturdy	[ˈstɜːdi]	adj.	坚固的,结实的
shaft	[ʃɑːft]	n.	轴;螺杆
collar	[ˈkɒlə(r)]	n.	箍;卡圈;套管;颈圈
gear	[gɪə(r)]	n.	齿轮
simultaneously	[ˌsɪməlˈteɪnɪəsli]	adv.	同时
multipoint tool			多刃刀具
engine lathe			普通车床
turret lathe			六角(转塔)车床
tracing and duplicating lathe			仿形车床
boring machine			镗床
drill machine			钻床
mill machine			铣床
broaching machine			拉床
sawing machine			锯床
gear-cutting machine			齿轮切割机床
cylindrical grinder			外圆磨床
centreless grinder			无心磨床
chemical milling			化学蚀刻(铣削)
electrical discharge machining			电火花加工
power feed			(机动力)进给
feed box			进给箱
T-slot			T型槽

The lathe bed is cast iron and provides accurately ground sliding surfaces(way) on which the carriage rides.

车身是铸铁件,它提供精确磨削过的滑动表面(导轨),上面放有拖板。(ground 是 grind 的过去分词,作定语用)

第 3 单元　车床

一台机床主要有以下三个功能:

(1) 固定工件或者刀架和刀具;

(2) 控制工件和刀具之间的相对运动;

(3) 提供一定范围的走刀和切削速度。

以去除切屑形式来加工金属的机床分类如下:

(1) 主要使用单点切削刀具的机床包括:普通车床、塔式车床、仿形车床、单轴自动车床、多轴自动车床、牛头刨床和龙门刨床、镗床。

(2) 使用多点切削刀具的机床包括:钻床、铣床、拉床、锯床、齿轮切割机床。

(3) 使用随机点切削刀具的机床包括:外圆磨床、无心磨床、平面磨床。

特殊金属切削方法包括:化学蚀刻铣削、电火花加工、超声波加工。

车床借助于转动的工件对着刀具来切去金属材料,以产生外圆柱面、内圆柱面或锥形表面。车床普遍靠端面切削来加工工件表面。在端面切削加工中,工件旋转,而刀具作垂直于回转轴线方向的移动。

普通车床,如图 3-1 所示,作为最基本的车床,是研制其他车床的基础。驱动电机装在床身底部并通过齿轮、皮带来驱动主轴。主轴是一根坚固的空心轴,装在重型轴承之间,其前端用来安装驱动盘(花盘),以便把确定的运动传到工件上。该驱动盘可借助螺纹、凸轮锁紧机构或借助一个螺纹垫圈和键固定在主轴上。

车床的床身是铸铁件,它提供精确磨削过的滑动表面(导轨),上面放有拖板。车床拖板是 H 型的铸件,刀具安装在拖板的刀架上。溜板箱装在拖板前面,内装有驱动齿轮,可以顺着导轨或横跨导轨移动刀具和拖板,以提供所希望的刀具运动。

拖板上面的小刀架能使刀的夹具旋转任意角度。小刀架上的手轮和丝杆可以使刀具作线性运动,另一个手轮和进给螺纹提供横向进给,使小刀架垂直于导轨移动。溜板箱中的齿轮可以在拖板沿着导轨和横跨导轨移动时提供动力进给。进给箱齿轮将运动传给拖板并控制刀具相对于工件的运动速度。由于进给箱的移动运动是由主轴齿轮驱动的,因此,进给量直接与主轴速度有关。进给箱齿轮传动机构也用于加工螺纹并能加工 4 扣~224 扣/英寸的螺纹。

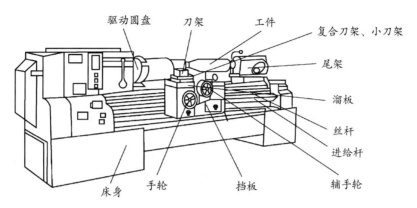

驱动圆盘　刀架　工件　复合刀架、小刀架　尾架　溜板　丝杆　进给杆　辅手轮　床身　手轮　挡板

图 3−1　普通车床

　　转塔车床基本上是具有某种附加特性的普通车床,用于半自动加工和减少人工操作误差。转塔车床的拖板设有 T 型槽,以便在车床导轨两侧安装夹刀装置。这样,当转塔转入合适位置时,可以正确地安装刀具,以便进行切削。拖板装有自动挡铁,以便控制刀具行程,提供良好的重复切削。转塔车床的尾座是六角形结构,可以装六把刀具。虽然操作前刀具和挡铁的安装要花大量时间,但一旦装刀完成,稍微熟练的工人就可以连续重复操作,直到刀具变钝需要更换为止。这样,为加工所做的准备时间相对于所制造的零件的数量是合理的时候,使用转塔车床在经济上才是可行的。

　　多轴自动车床装有 4、5、6 或 8 根主轴,在每根主轴中装一个工件。各主轴可以围绕着一根中心轴来转换位置,而主刀具溜板可以接近所有的主轴。每根轴位上都装有一侧向可以独立操作的刀具滑板。各刀具滑板都是靠凸轮操作的,因此,加工准备可能要花几天时间,至少 5 000 件的批量生产,它的使用才是合理的。这种机床的主要优点是所有的刀具能同时工作,因而一个工人可以看管几部机床。对于相对简单的零件,多轴自动车床可以在 5 秒内生产加工出一件产品。

READING MATERIAL

Machine Tools

Most of the mechanical operations are commonly performed on five basic machine tools:

(1) The drill press

(2) The lathe

(3) The shaper or planer

(4) The milling machine

(5) The grinder

Drilling is performed with a rotating tool called a drill. Most drilling in metal is done with a twist drill. The machine used for drilling is called a drill press.

Operations, such as reaming and tapping, are also classified as drilling. Reaming consists of removing a small amount of metal from a hole already drilled. Tapping is the process of cutting a thread inside a hole so that a cap screw or bolt may be threaded into it.

The lathe is commonly called the father of the entire machine tool family. For turning operations, the lathe uses a single-point cutting tool which removes metal as it travels past the revolving workpiece[1]. Turning operations are required to make many different cylindrical shapes, such as axes, gear blanks, pulleys, and threaded shafts. Boring operations are performed to enlarge, finish, and accurately locate holes.

Milling removes metal with a revolving, multiple cutting edge tool called milling cutter. Milling cutters are made in many styles and sizes. Some have as few as two cutting edges and others have 30 or more. Milling can produce flat or angled surfaces, grooves, slots, gear teeth, and other profile, depending on the shape of the cutters being used.

Shaping and planing produce flat surfaces with a single-point cutting tool. In shaping, the cutting tool on a shaper reciprocates or moves back and forth while the work is fed automatically towards the tool. In planing, the workpiece is attached to a worktable that reciprocates past the cutting tool. The cutting tool is automatically fed into the workpiece a small amount on each stroke.

Grinding makes use of abrasive particles to do the cutting. Grinding operations may be classified as precision or imprecision, depending on the purpose. Precision grinding is concerned with grinding to close tolerances and very smooth finish[2]. Imprecison grinding involves the removal of metal where accuracy is not important.

Words and Expressions

drill	[drɪl]	n.	钻头
		vt.	钻(孔)
mill	[mɪl]	n.	工厂;铣床
		vt.	铣削
twist	[twɪst]	n.	螺旋,麻花
press	[pres]	n.	压,按
ream	[riːm]	v.	铰孔
tap	[tæp]	v.	攻丝
screw	[skruː]	n.	螺钉
bolt	[bəʊlt]	n.	螺栓

axis	[æk'sɪz]	*n.*	轴
blank	[blæŋk]	*n.*	毛坯;坯料
pulley	[ˈpʊli]	*n.*	滑轮
bore	[bɔː]	*v.*	镗削
groove	[gruːv]	*n.*	槽口
profile	[ˈprəʊfaɪl]	*n.*	侧面
reciprocate	[rɪˈsɪprəkeɪt]	*vi.*	往复
tolerance	[ˈtɒlərəns]	*n.*	公差
be classified as			归类为
back and forth			来回
be attached to			与……相连接
feed... into			把……送进

NOTES

[1] For turning operations，the lathe uses a single-point cutting tool which removes metal as it travels past the revolving workpiece.

在车削加工中，车床利用单刃切削刀具对旋转的工件进行金属切除。[句中 which 引导的是一个定语从句，该从句修饰 cutting tool；as 引导的是时间状语从句，表示主句的动作（removes metal）与从句的动作（travels past）同时完成，it 指刀具]

[2] Precision grinding is concerned with grinding to close tolerances and very smooth finish.

精磨是为了缩小工件公差范围，以及降低表面粗糙度。

阅读材料

机　床

大多数机械加工操作通常是在以下五种基本机床上完成的：

(1) 钻床

(2) 车床

(3) 牛头刨床和龙门刨床

(4) 铣床

(5) 磨床

钻削是由旋转的钻头完成的。大多数金属的钻削由麻花钻来完成。用来进行钻削加工的机床称为钻床。扩孔和攻螺纹也归为钻削，扩孔是从已经钻好的孔上再切除少量的金属。攻螺纹是在内孔上加工出螺纹以使螺杆和螺栓拧进孔内。

车床通常被称为所有类型机床的始祖。在车削加工中,车床利用单刃切削刀具对旋转的工件进行金属切除。用车削可以加工各种圆柱体形状的工件,如轴、齿轮坯、带轮和丝杆轴。镗削可以将孔扩大,加工孔和精准定位孔。

铣削由旋转的、多切削刃的铣刀来完成。铣刀有多种类型和尺寸。有些铣刀只有两个切削刃,而有些则有多达三十或更多的切削刃。铣刀根据使用的刀具不同形状能形成平面、斜面、沟槽、狭槽、齿轮轮齿和其他外形轮廓。

牛头刨床和龙门刨床用单刃刀具来加工平面。用牛头刨床进行加工时,工件朝向刀具自动进给,刀具往复运动。在用龙门刨床进行加工时,工件安装在工作台上,工作台往复经过刀具而切除金属。刀具每完成一个行程则自动向工件进给一个小的进给量。

磨削利用磨粒来完成切削工作。根据加工要求,磨削可分为精密磨削和非精密磨削。精密磨削是为了缩小工件公差范围,以及降低表面粗糙度;在精度要求不高的地方切除多余的金属时,用非精密磨削。

Unit 4

Welding

Welding techniques have become so versatile that it is difficult nowadays to define "welding". Formerly welding was "the joining of metals by fusion," that is, by melting, but this definition will no longer do. Even though fusion methods are still the most common, they are not always used. Welding was next defined as the "joining of metals by heat," but this is no longer a proper definition either. Not only can metals be welded, but also many of the plastic can. Furthermore, several welding methods do not require heat. Every machinist is familiar with heatless welding. Cold pressure welding is a proper production welding method under some circumstances. Besides pressure welding, we can weld with sound and even with light from the famous laser. Faced with a diversity of welding methods that increase year by year, we must here adopt the following definition of welding: "Welding is the joining of metals and plastics by methods that do not employ fastening devices."

There is no uniform method of naming welding processes. Many processes are named according to the heat source or shielding method, but certain specialized processes are named after the type of joint produced. Examples are spot and butt welding. An overall classification can not take account of this because the same type of joint may be produced by a variety of processes. Spot welding may be done by electric resistance, arc, or electron-beam processes and butt welding by resistance, flash or any of a number of other methods.

There are a number of methods of joining metal articles together, depending on the type of metal and the strength of the joint which is required.

Soldering is the process of joining two metals by a third metal to be applied in the molten state. Solder consists of tin and lead, while bismuth and cadmium are often included to lower the melting point. One of the most important operations in soldering is that of cleaning the surface to be joined, this may be done by some acid cleaner. Soldering gives a satisfactory joint for light articles of steel, copper or brass, but the strength of a soldered joint is rather less than a joint which is brazed, riveted or welded. These methods of joining metal are normally adopted for strong permanent joints.

The simplest method of welding two pieces of metal together is known as pressure welding. The ends metal are heated to a white heat—for iron, the welding temperature should be about 1300 ℃—in a flame[1]. At this temperature the metal becomes plastic. The ends are then pressed or hammered together, and the joint is smoothed off. Care must be taken to ensure that the surfaces are thoroughly clean first, for dirt will weaken the weld. Moreover, the heating of iron or steel to a high temperature causes oxidation, and a film of oxide is formed on the heated surfaces. For this reason, a flux is applied to heated metal. At welding heat, the flux melts, and the oxide particles are dissolved in it together with any other impurities which may be present. The metal surfaces are pressed together, and the flux is squeezed out from the center of the weld. A number of different types of weld may be used, but for fairly thick bars of metal, a V-shaped weld should normally be employed. It is rather stronger than the ordinary butt weld.

The heat for fusion welding is generated in several ways, depending on the sort of metal which is being welded, and on its shape. An extremely hot flame can be produced from an oxyacetylene torch.

Generation of heat by an electric arc is one of the most efficient methods. Approximately 50% of the energy is liberated in the form of heat. The electric arc welding process makes use of the heat produced by the electric arc to fusion welded metallic pieces. This is one of the most widely used welding process, mainly because of the ease of use and high production rates that can be achieved economically. An arc is generated between two conductors of electricity, cathode and anode, when they are touched to establish the flow of current and then separated by small distance. An arc is sustained electric discharge through the ionized gas column called plasma between the two electrodes. It is generally believed that electrons liberated from the cathode move towards the anode and are accelerated in their movement. When they strike the anode at high velocity, large amount of heat is generated.

The endeavor of the welder is always to obtain a joint which is as strong as the welded metal itself and at the same time, the joint is as homogeneous as possible. To this end, the complete exclusion of oxygen and other gases to the detriment of the weld quality is very essential[2]. In manual metal welding, the use of stick electrodes does this job to some extent but not fully. In inert gas shielded arc welding process, a high pressure inert gas is flowing around the weld electrode while welding would physically displace all the atmospheric gases around the weld metal to fully protect it.

Resistance welding process is a fusion welding process where both heat and pressure are applied on the joint but no filler metal or flux is added. The heat necessary for the melting of the joint is obtained by the heating effect of the electrical resistance of the join and hence，the name resistance welding. In resistance welding，a low voltage(typically 1 V) with a very high current(typically 15 000 A) is passed through the joint for a very short time(typically 0.25 s). This high amperage heats the joint，due to the contact resistance at the joint and melts it. The pressure on the joint is continuously maintained and the metal is fused together under this pressure.

Words and Expressions

fusion	['fjuːʒən]	v.	熔接
shield	[ʃiːld]	v.	屏蔽,遮护,保护
butt welding			对接焊
solder	['səʊldə]	v.	软钎焊,锡焊
bismuth	['bɪzməθ]	n.	铋
cadmium	['kædmɪəm]	n.	镉
braze	[breɪz]	v.	硬钎焊,铜焊
rivet	['rɪvɪt]	v.	铆接
oxy-acetylene			氧乙炔
flux	[flʌks]	n.	焊剂
cathode	['kæθəʊd]	n.	阴极,负极;正极(原电池中)
ionize	['aɪənaɪz]	v.	电离,离子化
plasma	['plæzmə]	n.	等离子体

NOTES

〔1〕 The ends metal are heated to a white heat—for iron，the welding temperature should be about 1 300 ℃—in a flame.

火焰将金属的端部加热到白热状态——铁的焊接温度应在 1 300 ℃ 左右。(本句中 white heat 的状语 in a flame 被插入的说明语隔开了,也可另外表述为: The ends metal are heated to a white heat in a flame. The welding temperature for iron should be about 1 300 ℃.)

〔2〕 To this end，the complete exclusion of oxygen and other gases to the detriment of the weld quality is very essential.

为此,完全排除有害于焊接质量的氧和其他气体非常重要。(句中 To this end = For this purpose,介词短语 to the detriment of... 是 oxygen and other gases 的后置定语,意为"有害于……")

第四单元　焊接

　　焊接技术的应用如此广泛,以至于现在都很难定义"焊接"了。原先,焊接被定义为"通过熔融连接金属",也就是说,通过熔化,但这种定义已不再适用。尽管熔融法仍然是最普通的,却并非总被采用。焊接随后被定义为"借助热量将金属连接",现在这个定义也不完善。不仅金属可以焊接,很多塑料也可以,很多焊接方法并不需要热量。每个机械工人都对无热焊接很熟悉。冷压焊接在某些情况下是很合适的生产焊接方法。除压力焊接外,我们还可以用声波甚至激光来焊接。面对历年创造的多样化焊接方法,我们必须在这里采用下述焊接定义:"焊接是金属和塑料不采用紧固件的连接法。"

　　命名焊接没有一致的方法。许多焊接过程根据热源和保护方法来命名,某些专门的焊接过程按照形成的连接形式来命名,例如,点焊和对焊。总体的分类不可能考虑这一点,因为同样的连接可以通过许多焊接过程产生。点焊可借助电阻焊、电弧焊或电子束焊来完成,而对焊可以借助电阻焊、闪光焊(火花对焊)或其他多种方法中的一种来完成。

　　根据金属的种类和焊接所需的强度,可有几种将金属件连接到一起的方法。

　　软钎焊是通过施加第三种熔化金属来连接两种金属的方法。焊接材料有锡和铅,同时加入铋和镉来降低熔化点。软钎焊极重要的一项操作是对焊接表面的清洁,这可用某种酸来完成。软钎焊对轻型钢、紫铜或黄铜工件的焊接效果不错,但软钎焊的强度比硬钎焊、铆接或(常规)焊接低很多。这些连接金属的方法通常用于高强度的永久连接。

　　将两块金属焊接在一起的最简单的方法为压焊。火焰将金属的端部加热到白热状态——铁的焊接温度应在 $1\,300\,^{\circ}\mathrm{C}$ 左右。在这个温度下,金属成为塑性状态。然后将端部挤压或用锤敲打在一起,再将接头修平。应当注意确保先将焊接表面彻底清洁,否则污物会降低焊接强度。尤其是铁和钢加热到很高温度时会引起氧化,并在加热表面形成一层氧化膜。因此,要在加热表面施加焊剂。在焊接热作用下,焊剂融化,氧化物颗粒和任何其他可能存在的杂质熔解在焊剂里。金属表面被压在一起,焊剂被挤出焊接中心。可采用几种不同的焊缝,但对相当厚的金属条,一般使用 V 型焊缝,它比通常的平焊缝强得多。

　　按照被焊接金属的不同种类和形状,熔融焊接的热量可由多种方式产生。极热的火焰可由氧乙炔炬产生。

　　电弧生热是最高效的方法之一。大约 50% 能量以热能的形式释放。电弧焊利用电弧产生的热量来熔化被焊接的金属件。这是最广泛采用的焊接方法之一,主要是因为它易用且经济、高产。两个导电体,阴极和阳极接触会产生电流,然后分开微小距离,则在两极之间产生一道电弧。电弧是通过两极间称为等离子体的电离气柱的持续的放电产生的。一般认为从阴极释放的电子向阳极移动并在运动中加速。当它们高速撞击阳极时,就产生了大量的热。

焊接者的不懈努力总是要获得和被焊金属本身一样强度的连接,并且同时连接处要尽可能均匀。为此,完全排除有害于焊接质量的氧和其他气体非常重要。在手工电弧焊中,采用棒状电极可一定程度上完成这一任务。采用惰性气体保护焊过程中,焊接时一股高压的惰性气体从焊接电极周围流过,实际上排除了焊接金属周围的全部气体,从而起到保护作用。

　　电阻焊是一种在连接处同时施加热和压力的熔融焊接,但不采用填料金属或焊剂。熔化连接所需的热量通过连接的电阻效应获得,因而取名电阻焊。在电阻焊中,一股低电压(典型 1 伏)但很强的电流(典型 15 000 安)以很短的时间流经连接处(典型 0.25秒)。由于连接处的接触电阻,使这股强电流加热并熔化连接处。连接处的压力持续作用,使金属在压力作用下熔融在一起。

▉ READING MATERIAL

Machining

　　In addition to being familiar with the hardware components, to be an effective engineer one should also have some hands-on experience with the machine tools and machining technique that one day will be used to fabricate the products that you design[1].In this section, we discuss cutting with drill presses, band saws, lathes, grinders and mills, which are some of the tools available to perform standard machining operations in prototyping and production shops. Each of these tools is based on the principle of removing unwanted material from a workpiece through the cutting action of sharpened bits or blades.

　　A drill press is used to bore holes into a workpiece. A drill bit is held in the rotating chuck, and as a machinist turns the pilot wheel, the bit is lowered into the surface of the workpiece. The bit removes material in small chips and creates a hole in the workpiece. As should be the case, whenever metal is machined, the point where the bit cuts into the workpiece is lubricated. The oil reduces friction, and it also helps to remove heat from the cutting region. For safety reasons, vises and clamps securely hold the workpiece as the hole is bored in order to prevent material from shifting unexpectedly.

　　A band saw makes rough cuts through metal or plastic stock. The blade is a long, continuous loop having sharp teeth on one edge, and it rides on the drive and idle wheels. A variable-speed motor enables the operator to adjust the blade's speed depending on the type and thickness of material that will be cut. A tilting table is used for support. The machinist feeds the workpiece into the blade and guides it by hand, to make straight or slightly curved cuts. When the blade becomes dull and needs to be replaced, or if it should break, the band saw's internal blade grinder and

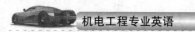

welder are used to clean up the blade's ends, connect them, and form a loop[2].

Grinding machine is a random point cutting tool which use abrasives in the shape of a wheel, bonded to a belt, a stick or simply suspended in liquid. On a cylindrical grinding machine the grinding wheel rotates between 5,500 and 6,500 r/min, while the work rotates between 60 and 125 r/min. The depth of cut is controlled by moving the wheel head. Coolants are provided to reduce heat distortion and to remove chips and abrasive dust. The grinding process is very important in production work for several reasons:

(1) It is the most common method for cutting hardened tool steel or other heat-treated steel.

(2) It can provide surface finish to 0.5 μm without high cost.

(3) The grinding operation can assure accurate dimensions in a relatively short time.

(4) Tiny and thin parts can only be finished by this method.

A lathe holds the workpiece and rotates it, and the sharpened edge of a tool cuts and removes material. Some applications of a lathe include the production of shafts, and thread cutting. The chuck on headstock hold one end of a workpiece, and the drive mechanism spins it rapidly about its centerline. The tailstock can be used to provide support to the otherwise free end of a long workpiece. As a cutting tool fed against the bar stock and moved along its length, the diameter of the workpiece is reduced to a desired dimension. Shoulders that will locate bearings on a shaft, grooves for holding retaining clips, and sharp changes in diameter of a stepped shaft can each be made in this manner.

Planing is a relatively simple cutting operation which flat surfaces, as well as various cross sections with grooves and notches, are produced along the length of the workpiece. Planing is usually done on large workpieces—as large as 25 m × 15 m. In a planer, the workpiece is mounted on a table that travels along a straight path. A horizontal cross-rail, which can be moved vertically, is equipped with one or more tool heads. The cutting tools are attached to the heads, and machining is done along a straight path. Because of the reciprocating motion of the workpiece, elapsed non-cutting time during the return stroke is significant.

A mill is versatile machine tool which is useful for machining the rough surfaces of a workpiece flat and smooth and for shaping them with slots, and holes. The workpiece is held by a vise on an adjustable table so that the part can be accurately moved in three directions(in the plane of the table and perpendicular to it)to locate the workpiece beneath the cutting bit. In a typical application, a piece of metal plate would be cut to approximate shape with a band saw, and the mill would be used to machine the surfaces and edges smooth, square, and to the final dimensions.

A machine component known as a lead screw is often used in the mill's internal mechanisms for positioning the work table beneath the spindle. The lead screw converts rotation of a shaft, produced either by a hand wheel or an electric motor, into the straight-line motion of the work table. As the lead screw turns and engages the nut(which is not allowed to rotate), the nut moves along the screw with each rotation by an amount equal to the distance between threads.

By using electric motors to drive lead screws and position the workpiece or cutting bit, mills and other machine tools can be controlled by computers. In that manner, a shop operation can be automated to achieve high precision or to complete a repetitive task on a large number of parts. A computer numerically controlled (CNC) mill performs the same types of operations as a conventional mill, but instead of being manually operated, it is programmed either through entry on a keypad or by downloading machining instructions that are created by computer-aided engineering software. CNC machine tools offer the potential to seamlessly produce physical hardware directly from a computer-generated drawing. With the ability to quickly reprogram machine tools, even a small general-purpose shop can produce a variety of high-quality machine components.

The fabrication techniques that an engineer selects for a certain product will depend, in part, on the time and expense of setting up the tooling. Some devices(for instance, air-conditioner compressors, microprocessors, hydraulic valves, and tires) are mass-produced, implying a process that is based on widespread mechanical automation. Mass production might be what you first imagine to take place in a large factory, and the manufacture of automobile engines is a prime example. The assembly line in that case comprises custom tools, fixtures, and specialized processes that are capable of producing only certain types of engines for certain vehicles. The assembly line will be set up so that only a small number of operations need to be performed in any one work area before the engine moves on to the next stage. Because engines will be produced at a relatively high rate, it is cost-effective for a company to allocate a large amount of factory floor space and many expensive machine tools to the production line, each of which might only drill a few holes or make a few welds[3]. Aside from hardware that is produced through mass manufacturing, other products are unique(for instance, the Hubble space telescope) or made in relatively small quantities (such as commercial jetliners), the best production method for a given product will ultimately depend on the quantity to be produced, the allowable cost, and the level of part-to-part variability that is acceptable.

Words and Expressions

machine tool			机床
drill press			钻床
pilot wheel			(操纵用)手轮
band saw			锯床,带锯
dull	[dʌl]	adj.	钝的
bore	[bɔ:]	v.	钻,钻孔
chuck	[tʃʌk]	n.	卡盘
lubricate	['lju:brɪkeɪt]	v.	润滑
vise	[vaɪs]	n.	台钳
stock	[stɒk]	n.	毛坯
loop	[lu:p]	n.	环,环路
tilt	[tɪlt]	v.	使倾斜
curve	[kɜ:v]	n.	曲线,弧线
coolant	['ku:lənt]	n.	冷却剂
plane	[pleɪn]	v.	刨削
elapse	[ɪ'læps]	v.	(时间)过去,消逝
spindle	['spɪndl]	n.	心轴,主轴,纺锤
air-conditioner		n.	空调机
compressor	[kəm'presə]	n.	压缩机
mass-produce			批量生产
Hubble space			哈勃天文望远镜
jetliner	['dʒetlaɪnə]	n.	喷气式客机

NOTES

［1］In addition to being familiar with... to fabricate the products that you design.

除熟悉机械零件外,作为一个实际的工程师还要具有一些机床和机加工技术的第一手实践经验,也许有一天这些经验将用于制造你所设计的产品。(这个长句主语是 one, In addition to 短语是动词的方式状语, to be an effective engineer 是目的状语。hands-on experience 是指某人对某事做过或运用过的经验,而不是仅仅读过或学到过的经验)

［2］When the blade becomes dull and needs to... and form a loop.

当锯条变钝或万一断裂需要更换时,锯床的内部磨齿机和焊机可用来清洁锯条的端部,将它连接成环状。(本句使用了并列主语 grinder and welder 和并列谓语 clean up,connect,form。如果具备专业素养,可以看出其结构本应是分列的,即用 grinder 来 clean up,用 welder 来 connect。两种工具联合作业,使 saw 重新 form a loop。为使语句不累赘,在意思不混淆的情况下采用了并列结构)

[3] Because engines will be produced at a... of which might only drill a few holes or make a few welds.

因为机器生产高效,成本效益高,所以一家公司可能会投入大量的厂房,为生产线购置昂贵的机械工具,也许这些机器仅需钻少数几个孔,或焊接几个地方。(组合词 cost-effective 意为"成本高效的"。本句 it 作形式主语,for 短语中的 machine tools 有一个很长的从句做补充说明,结构上把它放在后边)

■ 阅读材料

机加工

除熟悉机械零件外,作为一名工程师还要具有一些机床和机加工技术的第一手实践经验,也许有一天这些经验将用于制造你所设计的产品。本处我们讨论用钻床、带锯、车床、磨床和铣床切削,这些机床是样机加工和生产车间进行标准机械加工的可选用工具。每一种机床都是基于通过刀具切削作用去除工件多余材料的原理。

钻床在工件上钻孔。钻头被夹持在旋转的卡盘上,当机械工人转动手轮,钻头降低进入工件。钻头除去变成碎屑的材料,并在工件上打一个孔。正如实际情况那样,任何时候加工金属,在刀头切入工件的点上都要润滑。油液减少了摩擦力,也有助于从切削区带走热量。为了安全起见,钻孔时夹具牢牢地夹住工件,以防材料意外移位。

带锯粗糙地切穿金属和塑料坯。锯条是边缘开了铣齿的连续长环,它在主动轮和从动轮间运行。高速电机使操作者可根据材料的种类和切削的厚度调节锯条的速度。一个可倾斜的工作台用来支承。机械工人将工件进给到锯条上并用手调节切成直线或略带弧形。当锯条变钝或万一断裂需要更换时,锯床的内部磨齿机和焊机可用来清洁锯条的端部,将它连接成环状。

磨床是一种采用轮状、黏结在带上、棒状或仅仅悬浮在液体中的研磨材料进行随机点切削的机床。在圆柱磨床上,砂轮以 5 500 至 6 500 转/分的速度旋转,而工件以 60 至 125 转/分的速度旋转。切削深度通过移动砂轮头来控制。冷却剂被用来降低热翘曲并将切屑和磨料粉带走。磨削加工非常重要,原因如下:

1. 是切削硬质工具钢或其他热处理过的钢材的最普通的方法。

2. 无须高成本即可提供至 0.5 微米的表面精度。

3. 磨削加工可以在相对较短的时间内保证精确的尺寸。

4. 细小和薄的零件只能用这种方法精加工。

车床夹持工件并使它旋转,而锋利的刀具则切除材料。车床的应用包括轴的加工和螺纹切削。床头箱上的卡盘夹住工件的一端,传动机构使之快速地绕中心线旋转。

尾架可用来为长工件另一自由端提供支承。当刀具对着工件进给并沿长度方向移动时,工件的直径被减小到需要的尺寸。安装轴承的轴肩,容纳止推环的槽,阶梯轴的直径突变部位都可用这种方法制造。

刨削是相对简单的一种切削作业,用这种方法可沿工件的纵向加工出平面以及各种带沟作槽的截面。刨削一般用于大工件——大至25米×15米。在刨床上,工件被安装在直线移动的工作台上。可垂直移动的横向水平导轨上装备有一个或多个刀头。刀具安装在刀头上,切削沿纵向完成。由于工件的往复运动,返程花去的非切削时间很多。

铣床用于加工扁平工件的粗糙表面,在工件上打槽、孔,是一种很有用的多功能机床。工件由可调工作台上的台钳夹持,以便能够沿3个方向精确运动(工作台平面内和垂直于工作台方向)来定位刀具下的工件。在典型的应用中,一块钢板可用锯床锯成大致的形状,用铣床来加工表面和边缘,使之平整、方正,并达到最终尺寸。

通常在铣床的内部机构中常用一个称为丝杠的部件来将工件定位在主轴下。丝杠将手轮或电动机产生的轴向旋转运动转化为工作台的直线运动。当丝杠转动并结合螺母(不允许转动)时,螺母沿丝杠移动,丝杠每转一圈,螺母移动一个螺距。

在借助电动机来驱动丝杠并定位工件或刀具的情况下,铣床或其他机床可实现计算机控制。按这种方式,车间的作业可实现自动化,以达到高精度或在大量零件上完成重复的任务。尽管一台计算机数字控制(CNC)机床做的工作与普通铣床一样,但它不是手动操纵,而是由键盘输入的指令操控或由下载的工程软件生成的机器指令进行操纵的。数控机床提供从计算机图纸到硬件"无缝"生产的潜力。由于机床快速再编程的能力,甚至一个小型通用车间可生产大量不同的高质量机器零件。

工程师为某一特定产品选用的加工工艺将部分取决于调整加工工具的时间和成本。某些装置(例如,空调压缩机、微处理器、液压阀和轮胎)是批量生产的,这意味着一个基于广泛采用的机械自动化过程。批量生产可能是你最先想象到发生在大工厂的情况,汽车发动机生产就是一个典型的例子。这种情况下的装配线包括专用工具、夹具和仅能生产某种车上特定型号发动机的专门化工艺。装配线的建立使得任一个工位在发动机移动到下一工位前只需完成少量的操作。因为机器生产高效,成本效益高,所以一家公司可能会投入大量的厂房,为生产线购置昂贵的机械工具,也许这些机器仅需钻少数几个孔,或焊接几个地方。除了批量生产的硬件,其他产品无论是单件加工(例如,哈勃天文望远镜)还是少量生产(例如,商用喷气式客机),对某一给定产品的最佳生产方法将最终取决于生产的数量、成本开支范围和可接受的零件与零件之间的差异级别。

Unit 5

CAD and Applications

Before we present the basics of CAD, it is appropriate to give a brief history. CAD is a product of the computer era. It originated from early computer graphic systems to the development of interactive computer graphics. Such two systems include the Sage Project at the Massachusetts Institute of Technology (MIT) and Sketchpad. The Sage Project was aimed at developing CRT displays and operating systems. Sketchpad was developed under the Sage Project. A CRT display and light pen input were used to interact with the system. This coincidentally happened at about the same time that NC and APT (Automatically Programmed Tool) first appeared[1]. Later, X-Y plotters were used as the standard hard-copy output device for computer graphics. An interesting note is that an X-Y plotter has the same basic structure as a NC drilling machine except that a pen is substituted for the tool on NC spindle[2].

In the beginning, CAD systems were no more than graphics editor with some built-in design symbols. The geometry available to the user was limited to lines, circular arcs, and the combination of the two. The development of free-form curves and surfaces, such as Coon's patch, Bezier's patch, and B-spline, enables a CAD system to be used for sophisticated curves and surface design. Three-dimensional CAD systems allow a designer to move into the third dimension. Because a three-dimensional model contains enough information for NC cutter-path programming, the linkage between CAD and NC can be developed. So-called turnkey CAD/CAM systems were developed based on this concept and became popular in the 1970s and 1980s.

The 1970s marked the beginning of a new era in CAD—the invention of three-dimensional solid modeling. In the past, three-dimensional, wire-frame models represented an object only by its bounding edges. They are ambiguous in the sense that several interpretations might be possible for a single model. There is also no way to find the volumetric information of a model. Solid models contain complete information; therefore, not only can they be used to produce engineering drawing, but also engineering analysis can be performed on the same model as well. Later, many commercial systems and research systems were developed. Quite a few of these systems were based on the PADL and BUILD systems. Although they are powerful in representation, many deficiencies still exist. For example, such systems have extreme computation. It was in the mid-1980s that solid modelers made their way into the design environment. Today, their use is as common as drafting and wire-frame model applications.

CAD implementations on personal computers(PCs) have brought CAD to the masses. This development has made CAD available and affordable. CAD originally was a tool used only by aerospace and other major industrial corporation. The introduction of PC CAD packages, such as, AutoCAD, VersaCAD, CADKEY, and so on, has made it possible for small companies and even individuals to own and use CAD systems. By 1988, more than 100 000 PC CAD packages had been sold. Today, PC-based solid modelers are available and are becoming increasingly popular. Because rapid developments in microcomputers have enabled PCs to carry the heavy computational load necessary for solid modeling, many solid modelers now run on PCs and the platform has become less of an issue. With the standard graphics user interface(GUI), CAD systems can be ported easily from one computer to another. Most major CAD systems are able to run on a variety of platforms. There is little difference between mainframe, workstation, and PC-based CAD systems.

Words and Expressions

appropriate	[əˈprəʊprɪt]	adj.	适当的
graphic	[ˈgræfɪk]	adj.	绘画似的,图解的
interactive	[ɪntərˈæktɪv]	adj.	交互式的
Massachusetts	[mæsəˈtʃuːsɪts]	n.	马萨诸塞州(美国州名)
coincidental	[kəʊɪnsɪˈdentl]	adj.	一致的,符合的,巧合的
plotter	[plɒtə]	n.	绘图仪
geometry	[dʒɪˈɒmɪtri]	n.	几何学
arc	[ɑːk]	n.	弧,弓形,拱
curve	[kɜːv]	n.	曲线,弯曲,曲线图表
sophisticated	[səˈfɪstɪkeɪtɪd]	adj.	复杂的
dimension	[dɪˈmenʃən]	n.	尺寸,维(数)
cutter-path	[ˈkʌtəpɑːθ]	n.	刀具路径,切割轨道
linkage	[ˈlɪŋkɪdʒ]	n.	连接
concept	[ˈkɒnsept]	n.	观念,概念
ambiguous	[æmˈbɪgjʊəs]	adj.	暧昧的,不明确的

representation	[ˌreprɪzenˈteɪʃn]	*n.*	表示法，表现
deficiency	[dɪˈfɪʃənsi]	*n.*	缺乏，不足
extreme	[ɪksˈtriːm]	*adj.*	极端的，极大的
APT（Automatically Programmed Tool）			自动编程语言

NOTES

〔1〕This coincidentally happened at about the same time that NC and APT
（Automatically Programmed Tool）first appeared.

CAD 与初次出现的 NC 和 APT（自动编程工具）碰巧同时问世。（coincidentally
为插入语，at about the same time 为介词短语，可译为"几乎""同时"）

〔2〕An interesting note is that an X-Y plotter has the same basic structure as a
NC drilling machine except that a pen is substituted for the tool on NC spindle.

一个有趣的现象是 X-Y 绘图仪与 NC 钻床具有相同的基本机构，除了绘图笔被
NC 机床上的主轴刀具替代之外。（the same... as 可译为"与……相同"，as 为介词，引
导短语）

第 5 单元　CAD 及其应用

在讲述 CAD（计算机辅助技术）的基本理论之前，先说说它的简史是比较合适的。
CAD 是计算机时代的产品。它从早期的计算机绘图系统发展到现在的交互式计算机
图形学。两个这样的系统包括：麻省理工学院的 Sage Project 及 Sketchpad。Sage
Project 旨在开发 CRT 显示器及操作系统。Sketchpad 是在 Sage Project 下发展起来
的。CRT 显示器和光笔输入用于与系统进行交互操作。CAD 与初次出现的 NC（数字
控制）和 APT（自动编程工具）碰巧同时问世。后来，X-Y 绘图仪作为计算机绘图的标
准硬拷贝输出装置使用。一个有趣的现象是 X-Y 绘图仪与 NC 钻床具有相同的基本
结构，除了绘图笔被 NC 机床上的主轴刀具替代之外。

开始，CAD 系统仅仅是一个带有内置设计符号的绘图编辑器。供用户使用的几何
元素只有直线、圆弧以及两者的组合。自由曲线及其曲面的发展，如昆氏嵌面、贝塞尔
嵌面以及 B 样条曲线，使 CAD 系统可用于复杂曲线与曲面设计。三维 CAD 系统允许
设计者步入三维设计空间。一个三维模型包含了 NC 刀具路径编程所需要的足够信
息，因此，能够开发 CAD 与 NC 之间联系的系统。所谓交钥匙的 CAD/CAM（计算机
辅助制造）系统便是根据这一概念开发的，并从 20 世纪 70 年代至 80 年代流行起来。

20 世纪 70 年代，三维实体建模的发明标志着 CAD 一个新时代的开始。过去的三
维线框模型仅用其边界来表达一个物体。这在某种意义上是含糊的，一个简单的模型
可能有几种解释，同时也无法获得模型的体积信息。实体建模包含完整的信息，因此，

它们不仅可用于生成工程图,而且也可在同一模型上完成工程分析。后来,三维实体建模还开发了许多商业系统和研究系统。这些系统中相当多的是基于 PADL 和 BUILD系统。尽管它们在表达上是强有力的,但仍然存在许多缺陷。例如,这种系统要有极强的计算能力。直到 20 世纪 80 年代中期,实体建模开始介入设计环境。今天,实体建模的应用如同绘图和线框模型应用一样普遍。

在个人计算机上,CAD 已走向大众化。这种发展使 CAD 应用面更广且价格更合理。CAD 原本作为一种工具仅被航空和其他主要工业企业使用。AutoCAD、VersaCAD、CADKEY 等个人机 CAD 软件包的引入,使小型公司乃至个人可以拥有并使用 CAD 系统。到 1988 年为止,已销售 10 万个以上的个人机 CAD 软件包。今天,基于个人计算机的实体建模易于获得,并且越来越受欢迎。由于微型计算机的迅速发展,使得个人计算机能够承受实体建模需要的计算负荷,如今许多实体建模在个人机上运行,并且作为平台已不成为一个问题。随着标准图形用户界面(GUI)的发展,CAD 系统可以很容易地从一台计算机传送至另一台计算机,且大多数 CAD 系统都能够在不同平台运行。大型计算机、工作站和基于个人计算机的 CAD 系统之间几乎没有区别。

READING MATERIAL

CAM and Applications

When a design has frozen, manufacturing can begin. Computers have an important role to play in many aspects of production. Numerically controlled(NC) machine tools need a part program to define the components being made; computer techniques exist to assist, and in some cases virtually automate the generation of part programs. Modern shipbuilding fabricates structures from welded steel plates that are cut from a large steel sheet. Computer-controlled flame cutters are often used for this task and the computer is used to calculate the optimum layout of the components to minimize waste metal. Numerically controlled pipe-bending machines are able to operate directly from part programs generated by pipe-routing software.

Printed circuit board assembly can also be improved by computer methods. Quality is maintained by computer-controlled automatic test equipment that diagnoses faults in a particular board and rejects defective boards from the assembly line. Computers are used extensively to plot the artwork used to etch printed circuit boards and also to produce part programs for NC drilling machines.

One of the most important manufacturing function is stock and production control. If the original design is done on a computer, obtaining lists of material requirements is straightforward. Standard computer data processing methods are employed to organize the work flow and order components when required(Figure 5-1).

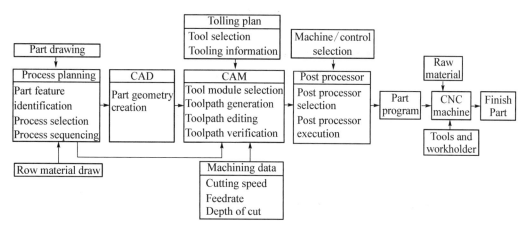

Figure 5 – 1   The block diagram of a CAD/CAM system

Part geometry requires calculation of a large number of tool positions. Part programming software is usually incorporated into a family of CAM (Computer Aided Manufacturing) software. Some CAM software is associated with CAD (Computer Aided Design) software into CAD/CAM stations. Then the CAM software can use the CAD files as a source of data, which speeds up the programming process.

Part programming software is user-friendly, meaning the programmer does not have to know the computer programming language or its operating system. It uses screen menus to lead the user through the programming process. Data can be entered via the keyboard, the mouse, or the function keys. Experienced programmers can use built-in macro capabilities and advanced techniques such as a family of parts to become even more productive.

Programming software has a dynamic graphics database to hold the actual machining sequences. These sequences can be viewed, edited, chained, or deleted. The programming can be accomplished whether single cuts or CNC machine canned cycle will be used. The software will also automatically calculate the proper feeds and speeds to be used during the machining, create a tooling list, and define the tool path.

Programmers can use different layers to associate with each profile being created or to construct clamps and fixtures to get a complete picture of the part setup. The tool motion can be seen as it will occur at the machine.

Using part programming software, the programmer can easily solve trigonometry problems to define an accurate tool path. When the program is done, the programmer can send it from the PC to the machine via a communication channel using built-in software with communications capability. Good part programming software is capable of:

（1）Establishing the machining parameters and tooling for a particular machine or job.

（2）Defining the geometry and tool path.

（3）Code generation，enabling the programmer to define what code is to be generated and how it is output to the machines.

（4）Communication enabling the programmer to use standard communications protocols or create his or her own.

Words and Expressions

numerically	[njuː'merɪkəli]	adv.	用数字,在数字上
define	[dɪ'faɪn]	v.	定义,详细说明
fabricate	['fæbrɪkeɪt]	vt.	制作,构成,捏造,伪造
optimum	['ɒptɪməm]	adj.	最适宜的
minimize	['mɪnɪmaɪz]	vt.	将……减到最少
diagnose	['daɪəgnəʊz]	v.	诊断
artwork	['ɑːtwɜːk]	n.	艺术品,美术品
etch	[etʃ]	v.	蚀刻
sequence	['siːkwəns]	n.	次序,顺序,序列
profile	['prəʊfaɪl]	n.	剖面,侧面,外形,轮廓
trigonometry	[trɪgə'nɒmɪtri]	n.	三角法
parameter	[pə'ræmɪtə]	n.	参数,参量,起限定作用的因素
protocol	['prəʊtəkɒl]	n.	草案,协议
optimum layout			最优设计
flame cutter			线切割
pipe-bending machine			折弯机
printed circuit board			印刷电路板
canned cycle			固定循环
canned data			存储数据
canned routine			固定程序

阅读材料

CAM(计算机辅助制造)及其应用

当设计确定之后,制造才能开始。计算机在生产的许多方面扮演着一个重要角色。数控机床需要编制一个零件加工程序来加工零件;计算机技术起到辅助作用,在某些情况下实质上是自动生成零件加工程序。现代造船是从大张钢板上切下焊接钢板来制造船体的。计算机控制的火焰切割机经常用于执行此项任务,而且计算机用于计算最佳排料,以使边角废料最少。数控弯管机能够通过管路软件生成的零件加工程序进行直接操作。

印刷电路板装配过程也可通过计算机方法加以改进。质量是由计算机控制的自动检测装置保证的,该装置能检测到某个板子上的缺陷,并且能从装配线上剔除有缺陷的板子。计算机广泛用于绘制蚀刻到印刷电路板的布线图,并且生成数控钻床所需的零件加工程序。

最重要的制造功能之一是库存和生产控制。如果原始设计是在计算机上进行的,可直接获取材料需求清单。标准的计算机数据处理方法可组织这项工作流程,并且按需订购零件(见图5-1所示)。

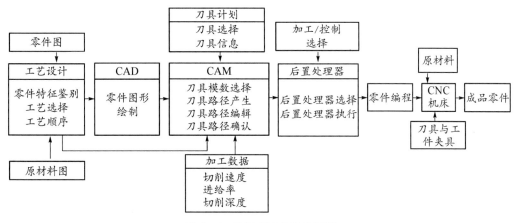

图 5-1　CAD/CAM 结构框图

零件的几何形状需要计算大量刀位。零件加工编程软件通常是并入一个计算机辅助制造软件包中的。一些计算机辅助制造软件与计算机辅助设计软件合并成计算机辅助设计与制造工作站。计算机辅助制造软件可使用计算机辅助设计文件作为数据源,这样加快了编程的进程。

零件加工编程软件是一个用户界面友好的软件,这意味着程序员不必懂得计算机编程语言或它的操作系统。它用屏幕菜单引导使用者完成编程过程。数据可通过键盘、鼠标或功能键输入。有经验的程序员可使用计算机巨大的内置容量和诸如系列零件簇的先进技术来获取更高的生产力。

编程软件有一个动态图形数据库来支持实际加工顺序。这些顺序可被显示、编辑、串联或删除。无论是单一切削,还是采用CNC(计算机数字控制)机床固定循环加工,程序都可生成。该软件也会自动计算加工中所用的适当进给量和切削速度,生成一个刀具清单和定义刀具路径。

程序员为生成的每一个轮廓分配不同的图层或用图层构建卡具与夹具来获得一个完整的零件工装图。可以看到与在机床上进行实际加工一样的刀具运动过程。

使用零件加工编程软件,程序员能轻松解决三角学问题,以确定准确的刀具路径。

当一个程序编好后,程序员可从个人计算机上通过通信线路用带有通信能力的内置软件将程序传送给数控机床。好的零件加工编程软件应具备如下条件:

1. 建立用于特定机床或任务的加工参数和刀具。
2. 定义几何模型及刀具路径。
3. 生成代码,能让程序员确定生成的代码以及如何将代码输出到机床。
4. 可使程序员使用标准的通信协议或生成他(她)自己的通信协议进行通信。

Unit 6

Computer Graphics

It is hard now to imagine a world without computer graphics. Today, graphic designers work almost exclusively with computer programs to create images once laboriously drawn with pen, compass, ruler, and T-square on paper[1]. Pilots learn the latest techniques of flying in flight simulators; engineers and architects design everything from aircraft to skyscrapers, making three-dimensional models with their computers; TV weather maps display precipitation as it occurs; and doctors can look inside a patient's body without breaking the skin. All these are because of the development of computer graphics.

Computer graphics is a sub-field of computer science and is concerned with digitally synthesizing and manipulating visual content. Although the term often refers to three-dimensional (3D) computer graphics, it also encompasses two-dimensional(2D) graphics and image processing.

2D computer graphics are mainly used in applications that were originally developed upon traditional printing and drawing technologies, such as cartography, technical drawing, advertising, etc. In those applications, the two-dimensional image is not just a representation of a real-world object, but an independent artifact with added semantic value; two-dimensional models are therefore preferred, because they give more direct control of the image than 3D computer graphics[2].

To create three-dimensional images, a computer uses a mathematical transformation called "projection". Although the images are presented on a two-dimensional screen, the computer, through using the principle of perspective and through its ability to make quick calculations, can portray the object so as to make it appear to the human eye as a three-dimensional object[3]. These techniques are the basis for computer-aided drafting and computer animation.

Working in three-dimensions with the computer, graphics will be produced easily in two- or three-dimensional modes depending upon the application. Creating two-dimensional graphics such as xy plots and schematics will be accomplished with CAD software. Two-dimensional geometric primitives such as circles and rectangles which serve as the generating geometry for cylinders and prisms are a part of two-dimensional software. Special applications, for example, dimensioning, generally utilize a two-dimensional view or series of two-dimensional views.

Three-dimensional geometric modeling involves wire-frame, surface, and solid models. A wire-frame model shows a series of nodes connected by lines to form an object. A surface model completely defines a series of areas connected to form the boundary of an object. The solid model is a total definition of an object which includes knowledge of all boundary and internal points. From a solid model, a complete analysis of performance of the object can be performed on the computer with appropriate software.

The engineering student of today will study graphics from the standpoint of supporting the design process. Geometric modeling techniques, analysis techniques which are mathematically based, and practice in visualization of three-dimensional geometries will be the focus of intensive computer utilization[4]. In order to prepare concepts for modeling and analysis, freehand techniques are still necessary to be studied and practiced. The student will learn to produce and interpret multi-views and pictorials both via sketches and computer techniques. Many of the graphics standards for appropriate representation of object features(sections, dimensioning, multi-views) will be studied.

Now, computer graphics has become a powerful design tool which promises to enhance significantly the engineer's ability to be creative and innovative in the solution of complex problems.

Words and Expressions

exclusively	[ɪkˈskluːsɪvli]	adv.	专门,仅仅,只
laboriously	[ləˈbɔːriəsli]	adv.	辛苦地
compass	[ˈkʌmpəs]	n.	圆规
precipitation	[prɪˌsɪpɪˈteɪʃn]	n.	降水,降水量(包括雨、雪、冰等);[化]沉淀
synthesize	[ˈsɪnθəsaɪz]	v.	合成;综合
manipulate	[məˈnɪpjuleɪt]	v.	操纵,操作
cartography	[kɑːˈtɒɡrəfi]	n.	地图制作,制图法,制图
artifact	[ˈɑːtɪfækt]	n.	人工制品,制造物
semantic	[sɪˈmæntɪk]	adj.	语义的
projection	[prəˈdʒekʃn]	n.	投影;投影图
perspective	[pəˈspektɪv]	n.	透视画法
animation	[ˌænɪˈmeɪʃn]	n.	卡通制作

schematic	[skiːˈmætɪk]	n.	框图
visualization	[ˌvɪzjuəlaɪˈzeɪʃn]	n.	显示,显像
pictorial	[pɪkˈtɔːriəl]	n.	图像

NOTES

[1] Today，graphic designers work almost exclusively with computer programs to create images once laboriously drawn with pen，compass，ruler，and T-square on paper.

如今,图形设计者几乎只用计算机程序来制作图像,这一工作曾经需要用铅笔、圆规、尺子和丁字尺很费力地在纸上来绘制完成。

[2] In those applications，the two-dimensional image is not just a representation of a real-world object，but an independent artifact with added semantic value；two-dimensional models are therefore preferred，because they give more direct control of the image than 3D computer graphics.

在那些应用中,二维图像不只是真实世界实物的展示,而且是具有另一重语义价值的独立制图;二维模型成为人们的首选,因为与三维计算机制图技术相比,它能对图像进行更多的直接控制。

[3] Although the images are presented on a two-dimensional screen，the computer，through using the principle of perspective and through its ability to make quick calculations，can portray the object so as to make it appear to the human eye as a three-dimensional object.

虽然图像呈现在二维屏幕上,但是运用透视画法原理及计算机能快速计算的功能,计算机能够描绘物体,使它在人的眼睛看来是三维的。

[4] Geometric modeling techniques，analysis techniques which are mathematically based，and practice in visualization of three-dimensional geometries will be the focus of intensive computer utilization.

几何建模技术,以数学为基础的分析技术和三维几何显像练习都是强化计算机应用的重心。

第6单元　计算机图形学

现在很难想象没有计算机图形学的世界会是什么样子。如今,图形设计者几乎只用计算机程序来制作图像,这一工作曾经需要用铅笔、圆规、尺子和丁字尺很费力地在纸上来绘制完成。飞行员在飞行模拟器上学习最新的飞行技术;工程师和建筑师设计从飞机到摩天大楼的一切物体,使用计算机制作三维模型;电视气象图演示预报犹如正在发生；医生还可以在不切破皮肤的情况下看到病人的身体内部。所有这些都得益于计算机图形学的发展。

计算机图形学是计算机科学的一个分支领域，它与数字合成和视觉图形的处理有关。虽然这个术语通常指的是三维（3D）计算机图形学，但也包括二维（2D）图形学和图像处理。

二维计算机图形学主要运用在从传统印刷和绘画技术发展起来的应用中，如地图制作、技术绘图、广告等。在那些应用中，二维图像不只是真实世界实物的展示，而且是具有另一重语义价值的独立制图；二维模型成为人们的首选，因为与三维计算机制图技术相比，它能更直接控制图像。

为创造三维图像，计算机使用了名为"投影"的数学变换法。虽然图像呈现在二维屏幕上，但是运用透视画法原理及计算机快速计算的功能，计算机能够描绘物体，使它在人的眼睛看来是三维的。这些技术是计算机辅助绘图和动画制作的基础。

用计算机进行三维制作，根据应用很容易就能得到二维或三维模式图形。用 CAD 软件可以创造二维图像，如 xy 坐标图和示意图。二维几何基本图形，如用以生成圆柱和棱柱的圆和长方形，就是二维软件的一部分。特殊应用，例如，尺寸标注，通常使用一个或一系列二维视图来表现。

三维几何建模包括结构线、面和实体模型。结构线模型给出一系列由线连接的点来构成物体，面模型可完整定义为由一系列连接起来的区域来构成物体的边界。实体模型是一个物体的总定义，它包括构成所有边界和内部点的知识。由实体模型可知，在计算机上用相关的软件就可以对物体性能做全面的分析。

今天的工程类学生要用支持设计过程的观点来学习图形学。几何建模技术，以数学为基础的分析技术和三维几何显像练习都是强化计算机应用的重心。为了建立建模和分析的观念，仍然有必要学习和练习徒手制图技术。学生要学习绘制并读懂多视图和图像，无论是手工绘制的还是计算机技术绘制的，还要学习物体特征的适当的表达方式（截面、尺寸标注、多视图）的一些图形标准。

目前，计算机图形学已经成为强大的设计工具，能极大地提高工程师解决复杂问题的创造和革新能力。

READING MATERIAL

Engineering Drawings

Engineering drawing is a graphical language used by engineers and other technical personnel associated with the engineering profession. The purpose of engineering drawing is to convey graphically the ideas and information necessary for the construction or analysis of machines, structures, or systems[1].

The basis for many engineering drawings is orthographic representation. Object are depicted by front, top, side, auxiliary, or oblique views, or combinations of these[2]. The complexity of an object determines the number of views shown. At times, pictorial views are also shown.

Engineering drawings often include such features as various types of lines, dimensions, lettered notes, sectional views, and symbols. They may be in the form of carefully planned and checked mechanical drawings, or they may be freehand sketches. Usually a sketch precedes the mechanical drawing.

Many objects have complicated interior detail which cannot be clearly shown by means of front, top, side, or pictorial views. Section views enable the engineer to show the interior detail in such cases. Features of section drawings are cutting-plane symbols, which show where imaginary cutting planes are passed to produce the sections, and section-lining, which appears in the section view on all portions that have been in contact with the cutting plane[3].

If the plane cuts entirely across the object, the section represented is known as a full section. If it cuts only halfway across a symmetrical object, the section is half section. A broken section is a partial one, which is used when less than a half section is needed. When only a part of the object is to be shown in section, conventional representation such as a revolved, rotated, or broken-out section is used. Thus, certain engineering drawings will be combinations of top and front views, section and rotated views, and partial or pictorial views.

In addition to describing the shape of objects, many drawings must show dimensions, so that workers can build the structure or fabricate parts that will fit together. This is accomplished by placing the required values along dimension lines. In metric dimensioning, the basic unit may be the meter, the centimeter, or the millimeter, depending upon the size of the object or structure.

Working types of drawings may differ in styles of dimensioning, lettering, positioning of the numbers, and in the type of fraction used[4]. If special precision is required, an upper and a lower allowable limit are shown. Such tolerance or limit dimensioning is necessary for the manufacture of interchangeable mating parts, but unnecessarily close tolerance are very expensive.

A set of working drawings usually includes detail drawings of all parts and an assembly drawing of the complete unit. Assembly drawings vary somewhat in character according to their use, as: design assemblies or layouts; working drawing assemblies; general assemblies; installation assemblies; and check assemblies[5]. A typical general assembly may include judicious use of sectioning and identification of each part with a numbered balloon. Accompanying such a drawing is a parts list, in which each part is listed by number and briefly described; the number of pieces required is stated and other pertinent information given[6]. Parts lists are best placed on separate sheets and typewritten to avoid time-consuming and costly hand lettering.

Words and Expressions

convey	[kən'veɪ]	v.	传达
orthographic	[ˌɔːθə'ɡræfɪk]	adj.	正交的,正射的
depict	[dɪ'pɪkt]	v.	描述,描绘
auxiliary	[ɔːɡ'zɪliəri]	adj.	附加的,辅助的
oblique	[ə'bliːk]	adj.	间接的,斜的
precede	[prɪ(ː)'siːd]	v.	在……之前
imaginary	[ɪ'mædʒɪnəri]	adj.	想象的,虚构的
revolve	[rɪ'vɒlv]	v.	旋转,循环
broken-out section			破断面
fabricate	['fæbrɪkeɪt]	v.	制造,建造,装配
metric	['metrɪk]	adj.	公制的,米制的
interchangeable	[ɪntə'tʃeɪndʒəbl]	adj.	可互换的
judicious	[dʒʊ(ː)'dɪʃəs]	adj.	贤明的,判断正确的
pertinent	['pɜːtɪnənt]	adj.	相关的,切题的

NOTES

[1] The purpose of engineering drawing is to convey graphically the ideas and information necessary for the construction or analysis of machines, structures, or systems.

工程制图的目的是以图形的方式传达机械、结构或系统的构造或分析所需的想法和信息。

[2] Object are depicted by front, top, side, auxiliary, or oblique views, or combinations of these.

物体由主视图、俯视图、侧视图、辅助视图、斜视图或者这几种视图的组合来描述。

[3] Features of section drawings are cutting-plane symbols, which show where imaginary cutting planes are passed to produce the sections, and section-lining, which appears in the section view on all portions that have been in contact with the cutting plane.

剖面图的特征有剖面符号和剖面线,剖面符号显示想象的剖面在哪里通过而产生剖视图,剖面线出现在剖视图中与剖面接触的所有部分。

[4] Working types of drawings may differ in styles of dimensioning, lettering, positioning of the numbers, and in the type of fraction used.

施工型图可以在尺寸标注、文字标注和数字标注位置的风格,以及所用的分数类型等上面有所差别。

[5] Assembly drawings vary somewhat in character according to their use，as：design assemblies or layouts；working drawing assemblies；general assemblies；installation assemblies；and check assemblies.

根据用途,不同装配图有所区别,如设计装配或布局、施工图装配、总装配、安装装配和检查装配。

[6] Accompanying such a drawing is a parts list，in which each part is listed by number and briefly described；the number of pieces required is stated and other pertinent information given.

这种图附有零件明细表,表中的每个零件按号码排列并加以简要描述,注明所需的件号并给出其他相关的信息。

■ 阅读材料

工程制图

工程制图是与工程专业相关的,由工程师及其他技术人员使用的一门图形语言。工程制图的目的是以图形的方式传达机械、结构或系统的构造或分析所需的想法和信息。

正视表示法是许多工程制图的基础。物体由主视图、俯视图、侧视图、辅助视图、斜视图或者这几种视图的组合来描述。物体的复杂程度决定了表达的视图的数量。有时也给出示意图。

工程制图通常包括这样的元素,如各种线型、尺寸、文字标注、剖视图和符号。它可以是经过详细计划和检查的机械制图的形式,或者是徒手画的草图的形式。在机械制图前通常是草图。

许多物体有复杂的内部细节,无法用主视图、俯视图、侧视图或示意图来清晰地表达。在这种情况下,剖视图可以让工程师表现内部细节。剖面图的特征有剖面符号和剖面线,剖面符号显示想象的剖面在哪里通过而产生剖视图,剖面线出现在剖视图中与剖面接触的所有部分。

如果剖面将物体完全切开,被表示的剖面叫作全剖视,如果仅切开对称物体的一半,则叫半剖视。局部剖视是部分剖视,在所需剖面少于半个剖面时使用。当只有物体的一部分需要用视图表达时,则使用传统的表达方法,如旋转剖视、转动视图或切面视图。这样,某些工程图就会由俯视图、主视图、剖视图和旋转视图以及局部视图或示意图组合而成。

除了描绘物体的形状,许多图必须标明尺寸,以便工人搭建物体结构或制作相配合的零部件,这一工作通过沿尺寸线标注所需数值来完成。在米制尺寸标注中,基本单位可以是米、厘米或毫米,这取决于物体或结构的尺寸。

施工型图可以在尺寸标注、文字标注和数字标注位置的风格，以及所用的分数类型等上面有所差别。如果需要特殊的精度，则要标明允许的上、下极限值。这种公差或极限的尺寸标注对于制造可以互换的相匹配的零部件是有必要的，但是没有必要的精密公差代价很高。

一组施工图通常包括所有部件的详图和整个装置的装配图。根据用途，不同装配图有所区别，如设计装配或布局、施工图装配、总装配、安装装配和检查装配。典型的总装配图包括剖切的正确使用和每个零件的识别，零件上有一个标有号码的圆圈。这种图附有零件明细表，表中的每个零件按号码排列并加以简要描述，注明所需的件号并给出其他相关的信息。零件明细表最好放在单独列明，并打印出来，以免费时费力地手写。

Unit 7

Computer Integrated Manufacturing System

Computer integrated manufacturing(CIM) is the term used to describe the modern approach to manufacturing. Although CIM encompasses many of the other advanced manufacturing technologies such as computer numerical control(CNC), computer-aided design/computer-aided manufacturing(CAD/CAM), robotics, and just-in-time delivery(JIT), it is more than a new technology or a new concept[1]. Computer integrated manufacturing is an entirely new approach to manufacturing, a new way of doing business.

The term computer integrated manufacturing was developed in 1974 by Joseph Harrington as the title of a book he wrote about tying islands of automation together through the use of computer[2]. It has taken many years for CIM to develop as a concept, but integrated manufacturing is not really new. In fact, integration is where manufacturing actually began. Manufacturing has evolved through four distinct stages:

- Manual manufacturing
- Mechanization/specialization
- Automation
- Integration

A CIM system is like an orchestra. Each component duty must work with the others through a single database, central software, and an open communications network. Each task is an independent process that must work with all others. The number of duties included in the CIM is chosen by the system managers and limited by the CIM software writers.

Each individual duty requires data and a RAM loaded with software to handle the transaction. This means that as a CIM unit becomes more capable(receives more duties) or has more users(more transaction per hour) or must contain more data, it must be a larger computer[3]. At a certain point, a single computer becomes incapable of handling the transactions and holding the necessary programming in RAM. Small CIM systems run on single computers. As more is asked of the CIM system, more equipment is needed.

A network of computers, which is controlled by a single central computer, can be developed. A major feature of a large CIM system is that each component of control is integrated through a single database. The database contains the store information about the products being produced. This data might be the part image from the CAD，or the records of the parts previously produced. This data is the record of the duties performed or about to be performed.

Single duties or groups of similar duties are assigned to smaller computers by the central master unit. Each smaller unit may share information through the central computer and its database.

Fully integrated manufacturing firms realize a number of benefits from CIM：

- Product quality increases
- Lead times are reduced
- Direct labor costs are reduced
- Product development times are reduced
- Inventories are reduced
- Overall productivity increases
- Design quality increases

Words and Expressions

integrated	[ˈɪntɪɡreɪtɪd]	adj.	集成的
approach	[əˈprəʊtʃ]	n.	途径,方法
encompass	[ɪnˈkʌmpəs]	vt.	包括,拥有
robotics	[rəʊˈbɒtɪks]	n.	机器人技术
just-in-time(JIT)			准时制造
evolve	[ɪˈvɒlv]	v.	进展,展开
distinct	[dɪˈstɪŋkt]	adj.	不同的,明显的
mechanization	[ˌmekənaɪˈzeɪʃn]	n.	机械化
orchestra	[ˈɔːkɪstrə]	n.	管弦乐队
transaction	[trænˈzækʃn]	n.	交易,处理
firm	[fɜːm]	n.	商行,公司
lead time			研制周期
inventory	[ˈɪnvəntri]	n.	详细目录,存货清单

NOTES

[1] Although CIM encompasses many of the other advanced manufacturing

technologies such as computer numerical control（CNC），computer-aided design/computer-aided manufacturing（CAD/CAM），robotics，and just-in-time delivery（JIT），it is more than a new technology or a new concept.

虽然 CIM 包含很多种其他的先进制造技术,如计算机数字控制(CNC)技术、计算机辅助设计/计算机辅助制造(CAD/CAM)技术、机器人技术以及准时制造技术,但它不仅仅是一项新技术或一个新概念。

［2］The term computer integrated manufacturing was developed in 1974 by Joseph Harrington as the title of a book he wrote about tying islands of automation together through the use of computer.

计算机集成制造这个术语来自约瑟夫·哈林顿 1974 年写的一本书的书名,在书中他描写了运用计算机将不同的自动化区域连接在一起的技术。

［3］This means that as a CIM unit becomes more capable（receives more duties）or has more users（more transaction per hour）or must contain more data，it must be a larger computer.

这意味着当 CIM 单元变得更强大(能接收更多的任务),或有更多的用户(每小时能处理更多的事务),或必须容纳更多的数据时,它必须是更大的计算机。

第 7 单元　计算机集成制造系统

计算机集成制造(CIM)是描绘现代制造方法的专业术语,虽然 CIM 包含很多种其他的先进制造技术,如计算机数字控制(CNC)技术、计算机辅助设计和计算机辅助制造(CAD/CAM)技术、机器人技术以及准时制造技术,但它不仅仅是一项新技术或一个新概念,还是一种全新的制造方法、一种新的业务经营途径。

CIM 这个术语来自约瑟夫·哈林顿 1974 年写的一本书的书名,在书中他描写了运用计算机将不同的自动化区域连接在一起的技术。CIM 观念的形成经历了很多年的时间,但是集成制造的概念并不新,事实上,集成就是制造的开始。制造的演变经历了 4 个明显的阶段:

- 手工制造
- 机械化/专业化
- 自动化
- 集成化

一个 CIM 系统就像一个管弦乐队,每个部件都必须通过同一个数据库、中央处理软件以及开放式的通信网络,并协调工作。每一项任务都是独立的,但必须和其余部分合作完成。CIM 系统执行的任务数目是由系统管理员选择的,而且受 CIM 软件编写器限制。

每一项任务的完成都需要数据和一个载有处理软件的随机存取存储器(RAM),这就意味着当一个 CIM 单元功能变得更强大(能接受更多的任务),或有更多的用户(每小时能处理更多的事务),或必须容纳更多数据时,它就必须是更大的计算机。从某种程度来说,单个计算机无法处理这样的任务并将所需程序保留在随机存取存储器(RAM)中。小型 CIM 系统可以在单个计算机上运行,当对 CIM 系统的要求越来越多时,对设备的需求也会随之增加。

由一个中央计算机进行控制,可以将多台计算机连接成一个网络,一个大 CIM 系统的主要特征是每个部件的控制集中到同一个数据库,该数据库存储了生产中产品的信息。这个数据可能是零部件 CAD 图像,或者是先前所生产的零部件的记录,是已执行或即将要进行的任务记录。

中央主处理单元将单项任务或一组类似的任务分配给小一点的计算机,每个小的单元可通过中央计算机及其数据库共享信息。

集成度很高的制造公司都知道 CIM 的很多优点:
- 提高了产品的质量
- 缩短了研制周期
- 减少了直接的劳动力成本
- 缩短了产品的研发时间
- 减少了存货清单
- 提高了总生产率
- 改善了设计质量

READING MATERIAL

Flexible Manufacturing Systems

In the modern manufacturing setting, flexibility is an important characteristic. It means that a manufacturing system is versatile and adaptable，while also capable of handling relatively high production runs. A flexible manufacturing system is versatile in that it can produce a variety of parts. It is adaptable because it can be quickly modified to produce a completely different line of parts.

A flexible manufacturing system（FMS）is a highly automated TG（group technology）machine cell，consisting of a group of processing workstations（usually CNC machine tools）, interconnected by an automated material handling and storage system，and controlled by a distributed computer system[1]. The reason the FMS is called flexible is that it is capable of processing a variety of different parts styles simultaneously at the various workstations，and the mix of patterns. The FMS is most suited for the mid-variety, mid-volume production range.

An FMS relies on the principles of group technology. Group technology is a manufacturing philosophy in which similar parts are identified and grouped together to take advantage of their similarities in design and production，while similar parts are arranged into part families，where each part family possesses similar design and manufacturing characteristics[2]. No manufacturing system can be completely flexible. There are limits to the range of parts or productions that can be made in an FMS. Accordingly，an FMS is designed to produce parts（or products）within a defined range of styles，sizes，and processes. In other words，an FMS is capable of producing a single part family or a limited range of part families.

An FMS must possess three capabilities：

● The ability to identify and distinguish among the different part or product styles processed by the system

　● Quick changeover of operating instructions

　● Quick changeover of physical setup

Flexible manufacturing systems can be distinguished according to the kinds of operations they perform：（1）processing operation or（2）assembly operations. An FMS is usually designed to perform one or the other but rarely both.

Flexible manufacturing systems can also be distinguished according to the number of machines in the system. Typical categories：（1）single machine cell，（2）flexible machine cell，and（3）flexible manufacturing system. Another classification of FMS is according to the level of flexibility designed into the system. This method of classification can be applied to systems with any number of workstations，but its application seems most common with FMCs（flexible manufacturing cell）and FMSs[3]. Two categories are distinguished here：dedicated FMS and random-order FMS.

An FMS has four basic components：

　● Workstations

　● Computer control system

　● Material handling and storage system

　● Human operators

Words and Expressions

characteristic	[ˌkærɪktəˈrɪstɪk]	n.	特性,特征,特色
versatile	[ˈvɜːsətaɪl]	adj.	多方面的,通用的
adaptable	[əˈdæptəbl]	adj.	能适应的,可修改的
modify	[ˈmɒdɪfaɪ]	v.	修改,更正
group technology			成组技术
interconnect	[ˌɪntə(ː)kəˈnekt]	vt.	使互相连接
simultaneously	[ˌsɪməlˈteɪnɪəsli]	adv.	同时地

philosophy	[fəˈlɒsəfi]	n.	哲学,基本定律,原理,原则 (思想)体系
similarity	[ˌsɪməˈlærəti]	n.	相似,类似
distinguish	[dɪsˈtɪŋgwɪʃ]	v.	区别,辨别
changeover	[ˈtʃeɪndʒəʊvə]	n.	转换
classification	[ˌklæsɪfɪˈkeɪʃn]	n.	分类,分级
dedicated	[ˈdedɪkeɪtɪd]	adj.	专用的
random-order			随机顺序,随机位,任意顺序

NOTES

[1] A flexible manufacturing system(FMS) is a highly automated TG(group technology) machine cell, consisting of a group of processing workstations(usually CNC machine tools), interconnected by an automated material handling and storage system, and controlled by a distributed computer system.

一个柔性制造系统(FMS)是高度自动化的 TG(成组技术)机床单元,包括一组加工产品的工作站(通常为计算机数字控制机床),与自动化的物料装卸和存储系统相连,并由分布式的计算机系统进行控制。

[2] Group technology is a manufacturing philosophy in which similar parts are identified and grouped together to take advantage of their similarities in design and production, while similar parts are arranged into part families, where each part family possesses similar design and manufacturing characteristics.

成组技术是一种制造体系,它先对类似的零件进行鉴别,利用它们在设计和生产上的相似性将它们归为一组,零件归到不同的零件组后,每一组零件都有相似的设计和加工制造特征。

[3] This method of classification can be applied to systems with any number of workstations, but its application seems most common with FMCs (flexible manufacturing cell) and FMSs.

这一分类方法可以用于有任意个工作站的系统,但是在应用中,似乎最常用的名称为 FMCs(柔性制造单元)和 FMSs。

阅读材料

柔性制造系统

在现代制造模式中,柔性是很重要的特性。柔性制造系统是一个多功用、可变通的制造系统,能适应相对较快的生产运行速度。一个柔性制造系统是多功用的,因为它可以生产多种零件;它可变通是因为它可以快速修改,以组成完全不同的零件生产线。

一个柔性制造系统(FMS)是高度自动化的 TG(成组技术)机床单元,包括一组加工产品的工作站(通常为计算机数字控制机床),与自动化的物料装卸和存储系统相连,并由分布式的计算机系统进行控制。FMS 系统被称作柔性的原因是它能同时在各个工作站上加工很多种不同风格的零件,且样式各异。FMS 系统最适合生产种类和产量都适中的产品。

FMS 系统以成组技术为原则。成组技术是一种制造体系,它先对类似的零件进行鉴别,利用它们在设计和生产上的相似性将它们归为一组,零件归到不同的零件组后,每一组零件都有类似的设计和加工制造特征。没有哪个制造系统是完全柔性的,能够在某一 FMS 系统进行加工的零件或产品的范围也是有限的。因此,一个 FMS 系统可设计成在生产零件(或产品)前确定样式、尺寸和加工程序,换句话说,一个 FMS 系统可以生产一组零件或有限组的零件。

一个 FMS 系统必须具有三种功能:

● 能识别和区分该系统生产的不同零件或产品类型;

● 操作指令的快速转换;

● 硬件的快速转换。

柔性制造系统可以由它们所执行的操作类型来加以区分:①加工操作或②组装操作。一个 FMS 系统通常设计成可执行其中一项功能,但很少两者兼备。

柔性制造系统还可以根据系统所包含的机床数目来分类。典型分类如下:① 单机床单元,② 柔性机床单元,以及③柔性制造系统。FMS 系统的另一个分类方法是根据设计赋予系统的柔性程度来分,这一分类方法可以用于有任意个工作站的系统,但是在应用中,似乎最常用的名称为 FMCs(柔性制造单元)和(FMSs)。在这里将 FMS 系统分为两类:专用 FMS 系统和随机顺序 FMS 系统。

一个 FMS 系统有四个基本部分:

● 工作站

● 计算机控制系统

● 物料装卸和存储系统

● 操作人员

Unit 8

The History of CNC and NC Development

扫码可见本项目
参考资料

One of the most fundamental concepts in the area of advanced manufacturing technologies is numerical control(NC). Prior to the advent of NC, all machine tools were manually operated and controlled. With manual control, the quality of the product is directly related to and limited to the skills of the operator.

The development of NC owns much to the United States Air Force and the early aerospace industry in 1940s. The first development work in the area of NC is attributed to John Parsons and his associate Frank Stulen at Parsons Corporation in Traverse City, Michigan. Parsons was a contractor for Air Force during the 1940s and had experimented with the concept of using coordinate position data contained on punched cards to define and machine the surface contours of airfoil shape[1]. He had named his system the Cardamatic milling machine, since the numerical data was stored on the punched cards.

The first NC machine was developed in 1952. Under contract to the US Air Force, the Parsons Corporation undertook the development of a flexible, dynamic manufacturing system, designed to maximize productivity by emphasizing details required to achieve desired accuracies[2]. This system would allow design changes without costly modifications to tools and fixtures, and it would fit into a modern, productive manufacturing management for small-to-medium sized production runs[3]. The Parsons Corporation subcontracted the development of the control system to the Massachusetts Institute of Technology (MIT) in 1951. MIT met the challenge successfully, and in 1952 demonstrated a Cincinnati Hydrotel milling machine equipped with the new technology, which was named Numerical control(NC) and used a pre-punched tape as the input media. Since 1952, practically every machine tool manufacturer in the Western world has converted part or its entire product to NC.

The first NC machines used vacuum tubes, electrical relays, and complicated machine-control interfaces(1952). It did not contain any actual central processing unit. The second generation of machines utilized improved miniature electronic tubes (1959), and later small-scale integrated circuits(1965).

As computer technology improved, NC underwent one of the most rapid changes known in history. The fourth generation used much improved integrated circuit(1970s). Computer hardware became progressive, less expensive and more reliable. And NC control builders introduced for the first time Read Only Memory (ROM) technology. ROM was typically used for program storage in special-purpose applications, leading to the appearance of the computer numerical control(CNC) system. CNC was successfully introduced to practically every manufacturing process.

The fifth generation is microprocessor CNC(MCNC). Since the introduction of NC in 1952, there have been dramatic advances in digital computer technology. The physical size and cost of a digital computer have been substantially reduced at the same time that its computational capabilities have been substantially increased. It was logical for the makers of NC equipment to incorporate these advances in computer technology into their products, starting first with microcomputer in the 1980s[4]. Among the strengths of the fifth generation microprocessor CNC(MCNC) are added part program memory storage, reduction of printed circuit boards, programmable interface, faster memory access, parametric subroutines, and macro capabilities[5].

With the advances in electronics and computer technology, current CNC systems employed several high-performance microprocessors and programmable logical controllers that work in a parallel and coordinated fashion[6]. Today, almost all the new machine control units are based on computer technology.

Words and Expressions

prior to			在……之前
advent	[ˈædvənt]	n.	出现,到来
aerospace	[ˈeərəuspeɪs]	n.	航空宇宙
associate	[əˈsəusieɪt]	n.	同伴,伙伴
contractor	[kənˈtræktə]	n.	立契约的人,承包商
punched card			穿孔卡,打孔卡片
airfoil	[ˈeəfɔil]	n.	翼剖面(翼型,方向舵)
modification	[ˌmɒdɪfɪˈkeɪʃn]	n.	修正,修改
subcontract	[sʌbkənˈtrækt]	n.	分契,转包合同
vacuum tube			真空管,电子管
small-scale	[ˌsmɔːlˈskeɪl]	adj.	小规模的
ROM(Read Only Memory)	[rɒm]	n.	只读存储器
microprocessor	[ˌmaɪkrəuˈprəusesə(r)]	n.	微处理器
substantially	[səbˈstænʃəli]	adv.	实质上,本质上

parametric	[ˌpærəˈmetrɪk]	*adj*. 参数的
subroutine	[ˈsʌbruːtiːn]	*n*. 子程序
macro	[ˈmækrəʊ]	*n*. 宏,巨

 **NOTES**

[1] Parsons was a contractor for Air Force during the 1940s and had experimented with the concept of using coordinate position data contained on punched cards to define and machine the surface contours of airfoil shape.

20 世纪 40 年代,帕森斯是空军的一个承包商,曾经尝试运用打孔卡片上包含的坐标位置数据来定义并加工飞机的翼剖面轮廓。

[2] Under contract to the US Air Force, the Parsons Corporation undertook the development of a flexible, dynamic manufacturing system, designed to maximize productivity by emphasizing details required to achieve desired accuracies.

按照与美国空军签订的合同,帕森斯公司承担了此柔性、动态制造系统的研制工作,通过强调细节设计,以达到所期望的精确度,将生产率最大化。

[3] This system would allow design changes without costly modifications to tools and fixtures, and it would fit into a modern, productive manufacturing management for small-to-medium sized production runs.

这个系统可以在对刀具和夹具不做重大修改的情况下对设计进行修改,而且能适应中小型生产的现代、多产的制造管理模式。

[4] It was logical for the makers of NC equipment to incorporate these advances in computer technology into their products, starting first with microcomputer in the 1980s.

对 NC 设备的制造者来说,将这些先进的计算机技术融入自己的产品是很自然的事,于是在 20 世纪 80 年代出现了微型计算机。

[5] Among the strengths of the fifth generation microprocessor CNC(MCNC) are added part program memory storage, reduction of printed circuit boards, programmable interface, faster memory access, parametric subroutines, and macro capabilities.

第五代数控机床功能强大,微处理器 CNC 增添了零件程序的内存存储,减少了印刷电路板,具有可编程功能的接口,内存的存取速度加快,子程序参数化,并具有宏的功能。

[6] With the advances in electronics and computer technology, current CNC systems employed several high-performance microprocessors and programmable logical controllers that work in a parallel and coordinated fashion.

随着电子和计算机技术的发展,当前的 CNC 系统使用若干高性能微处理器和可编程逻辑控制器并行协调工作。

第 8 单元　计算机数字控制(CNC)和数字控制(NC)的发展史

数字控制(NC)是先进制造技术领域最基本的概念之一。在 NC 技术出现之前,所有的机床都靠手动操作控制,在手动控制情况下,产品的质量与操作人员的技术直接相关,因此受到限制。

数字控制的发展与 20 世纪 40 年代的美国空军以及早期的航空工业有很大关系。NC 领域最初的发展工作归功于密歇根州 Traverse 城帕森斯公司的 John Parsons 和他的同伴 Frank Stulen。20 世纪 40 年代,帕森斯是空军的一个承包商,曾经尝试运用打孔卡片上包含的坐标位置数据来定义并加工飞机的翼剖面轮廓。他将系统命名为 Cardamatic 铣床,因为其数据是存储在打孔卡片上的。

第一台 NC 机床生产于 1952 年,按照与美国空军签订的合同,帕森斯公司承担了柔性、动态制造系统的研制工作,通过强调细节设计,以达到所期望的精确度,将生产率最大化。这个系统可以在对刀具和夹具不做重大修改的情况下对设计进行修改,而且能适应中小型生产的现代、多产的制造管理模式。1951 年,帕森斯公司将控制系统的研制转包给了麻省理工学院(MIT)。麻省理工学院出色地完成了这一挑战,并于 1952 年展出了蕴含新技术的 Cincinnati Hydrotel 铣床,命名为数字控制(NC),使用穿孔带作为输入介质。1952 年以后,几乎所有的西方机床制造商都将其产品部分或全部变成了数字控制。

最初的 NC 机床使用真空电子管、电力继电器和复杂的机械控制接口(1952 年),而不包含任何实质性的中央处理单元。第二代机床使用改进的微型电子管(1959 年),随后发展成小规模的集成电路(1965 年)。

随着计算机技术的发展,NC 经历了有史以来最快的改变,第四代 NC 机床运用了更先进的集成电路技术(20 世纪 70 年代)。计算机硬件变得先进、廉价且更加可靠,而 NC 控制系统的制造者第一次引入了只读存储器(ROM)技术。只读存储器的典型应用是在一些特殊应用中存储程序,因而导致了计算机数字技术(CNC)系统的出现,随后 CNC 系统被成功引入到每一个制造过程。

第五代数控机床是微处理器计算机数字控制(MCNC)。1952 年由于数字控制技术的引入,数字计算机技术发生了飞跃性的发展,一台数字计算机的实际尺寸大大缩小,同时造价也大大降低,而其计算能力却得到了巨大的提升。对 NC 设备的制造者来说,将这些先进的计算机技术融入他们的产品是很自然的事,于是在 20 世纪 80 年代出现了微型计算机。第五代数控机床功能强大,微处理器 CNC 增添了零件程序的内存存储,减少了印制电路板,具有可编程功能的接口,内存的存取速度加快,子程序参数化,并具有宏的功能。

随着电子和计算机技术的发展,当前的 CNC 系统使用若干高性能微处理器和可编程逻辑控制器并进行协调工作。现如今,几乎所有新机械设备的控制单元都以计算机技术为基础。

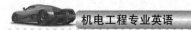

READING MATERIAL

The Advantages and Disadvantages of CNC Machines

CNC(Computer Numerical Control) machines are widely used in manufacturing industry. Traditional machines such as vertical millers, centre lathes, shaping machines, routers, etc., operated by a trained engineer have, in many cases, been replaced by computer control machines[1].

Advantages:

(1) CNC machines can be used continuously 24 hours a day, 365 days a year and only need to be switched off for occasional maintenance.

(2) CNC machines are programmed with a design which can then be manufactured hundreds or even thousands of times. Each manufactured product will be exactly the same.

(3) Less skilled people can operate CNCs unlike manual machines, which need skilled engineers.

(4) CNC machines can be updated by improving the software used to drive the machines.

(5) Training in the use of CNCs is available through the use of "virtual software". This is software that allows the operator to practice using the CNC machine on the screen of a computer. It is similar to a computer game.

(6) CNC machines can be programmed by advanced design software, enabling the manufacture of products that cannot be made by manual machines, even those used by skilled engineers[2].

(7) Modern design software allows the designer to simulate the manufacture of his/her idea. There is no need to make a prototype or a model. This saves time and money.

(8) One person can supervise many CNC machines. Once they are programmed they can usually be left to work by themselves. Sometimes only the cutting tools need replacing occasionally.

(9) A skilled engineer can make the same component many times. However, if each component is carefully studied, each one will vary slightly. A CNC machine will manufacture each component as an exact match.

Disadvantages:

(1) CNC machines are more expensive than manually operated machines, although costs are slowly coming down.

(2) The CNC machine operator only needs basic training and skills, enough to

supervise several machines. In years gone by, engineers needed years of training to operate centre lathes, milling machines and other manually operated machines[3]. This means many of the old skills are being lost.

(3) Fewer workers are required to operate CNC machines compared to manually operated machines. Investment in CNC machines can lead to unemployment.

(4) Many countries no longer teach students how to use manually operated machines. Students no longer develop the detailed skills required by engineers from the past. These include mathematical and engineering skills.

Obviously, CNC machines have more advantages than disadvantages. The companies that adopt CNC technology are increasing their competitive edge. In the future the broader use of CNC machines will be one of the best ways to enhance automation in manufacturing.

Words and Expressions

vertical	[ˈvɜːtɪkl]	adj.	垂直线;垂直的
router	[ˈruːtə]	n.	刻模机
maintenance	[ˈmeɪntənəns]	n.	维护,维修
virtual software			虚拟软件
simulate	[ˈsɪmjʊleɪt]	v.	模仿,模拟
prototype	[ˈprəʊtətaɪp]	n.	原型
supervise	[ˈsjuːpəvaɪz]	v.	监督,管理,指导
slightly	[ˌslaɪtli]	adv.	略微
unemployment	[ˌʌnɪmˈplɔɪmənt]	n.	失业
competitive	[kəmˈpetətɪv]	adj.	竞争的,比赛的
edge	[edʒ]	n.	边缘,优势
enhance	[ɪnˈhɑːns]	v.	提高,加强,增加

NOTES

[1] Traditional machines such as vertical millers, centre lathes, shaping machines, routers, etc., operated by a trained engineer have, in many cases, been replaced by computer control machines.

由训练有素的工程师操作的传统机床,如立式铣床、普通车床、牛头刨床、刻模机等,在许多情况下,已经由计算机控制的机床代替了。

[2] CNC machines can be programmed by advanced design software, enabling

the manufacture of products that cannot be made by manual machines, even those used by skilled engineers.

计算机数字控制机床可以用先进的设计软件来编程,因而可以加工手动机床无法完成的产品,甚至是很熟练的工程师操作也完成不了的产品。

[3] In years gone by, engineers needed years of training to operate centre lathes, milling machines and other manually operated machines.

过去,操作普通车床、铣床和其他手动操作的机床,工程师都需要接受几年的培训。

阅读材料

计算机数字控制(CNC)机床的优缺点

计算机数字控制(CNC)机床在制造工业领域有广泛应用。由训练有素的工程师操作的传统机床,如立式铣床、普通车床、牛头刨床、刻模机等,在许多情况下,已经由计算机控制的机床代替了。

优点:

1. CNC机床可以一天24小时、一年365天连续使用,只需要在做维护时偶尔断电;

2. CNC机床可以将一个设计编成程序,然后进行成百上千次的重复加工,每次加工出来的产品都会是一模一样的;

3. 技术不熟练的人也可以操作CNC机床,而手动机床不同,需要有技术的工程师来操作;

4. CNC机床可以通过改进驱动软件来进行更新;

5. CNC机床的训练可以通过"虚拟软件"来完成,操作者能在计算机屏幕上对CNC机床进行练习,类似于玩计算机游戏;

6. CNC机床可以用先进的设计软件来编程,因而可以加工手动机床无法完成的产品,甚至是很熟练的工程师操作也完成不了的产品;

7. 现代设计软件使得设计者可以按照其设计方案进行模拟加工,不需要制作原型或模型,这样节省了时间和金钱;

8. 一个人可以同时管理多台CNC机床,一旦编好了程序,通常它们就可以自行工作,无须看管,只是偶尔需要更换一下切削刀具;

9. 熟练的工程师可以多次重复制作出同样的零部件,然而,如果仔细研究一下,就能发现每一个都有些微的不同,但是CNC机床加工出来的每个零部件都是完全相同的。

缺点:

1. 虽然CNC机床的价格正在缓慢下降,但仍比手动操作机床贵得多。

2. CNC机床操作员只需要基本的训练和技能,就足够管理几台机床了。过去,操作普通车床、铣床和其他手动操作的机床,工程师都需要接受几年的培训,这意味着许多老技术正在丢失。

3. 与手动操作的机床相比,CNC 机床需要的操作工人更少,所以使用 CNC 机床会导致失业。

4. 许多国家都不再教学生怎样使用手动操作机床,学生也不再学习过去的工程师所需要的详细技能,包括数学技能和工程技能。

很显然,CNC 机床的优点多于缺点,采用 CNC 技术的公司都在增强自己的竞争优势。将来,CNC 机床的进一步广泛应用将会是提高制造自动化程度的最好方法。

Unit 9

CNC Systems

Current CNC systems allow simultaneous servo position and velocity control of all axis, monitoring of controller and machine tool performance, online programming with graphical assistance, in-process cutting process monitoring, and in-process part gauging for completely unmanned machining operations[1]. Manufacturers offer most of these features as options.

A typical CNC machine tool has five fundamental units: (1) the input media, (2) the machine control unit, (3) the servo-drive unit, (4) the feedback transducer, and (5) the mechanical machine tool unit. The general relationship among the five components is illustrated in Figure 9 - 1.

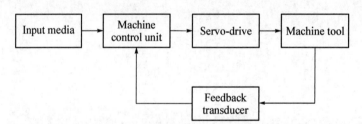

Figure 9 - 1　Basic components of a CNC machine tool

The input media contains the control program, also called the part program. It is the detailed step-by-step commands that direct the actions of the machine tool. The individual commands refer to positions of a cutting tool relative to the worktable on which the work-part is fixed. Additional instructions are usually included, such as spindle speed, feed rate, cutting tool selection, and other functions[2].

The program is coded on a suitable medium for submission to the machine control unit. The most common input medium on early NC systems was punched tape. Today, punched tape has largely been replaced by newer storage technologies in modern machine shops. A more advanced method is through a direct link with a computer. This is called direct numerical control(DNC). With DNC, a group of NC or CNC machines can be controlled simultaneously by a host computer.

The centerpiece of the CNC system is the control. The machine control unit （MCU） consists of a microcomputer and related control hardware that stores the program of instructions and executes it by converting each command into mechanical actions of machine tool，one command at a time[3].

Proprietary controls have a closed architecture. The systems are custom-built by the manufacturer and often contain closely guarded circuits，algorithms，and control programs. This type of control is expensive but offers very high reliability. In recent years，there has been a push toward so-called open architecture controls—that is，controls made from commonly available components and software. To some control builders this has meant PC-based controls.

In servo-drive unit，the drives in machine tools are classified as spindle and feed drive mechanisms. Spindle and feed drive motors and their servo-amplifiers are the components of the servo-drive unit. The MCU processes the data and generates discrete numerical position commands for each feed drive and velocity command for the spindle drive. The numerical commands are converted into signal voltage by the MCU unit and sent to servo-amplifiers，which process and amplify them to the high voltage levels required by the drive motors[4].

The function of a feedback system is to provide the control with information about the status of the motion control system. As the drives move，sensors measure their actual position. The difference between the required position and the actual position is detected by comparison circuit and the action is taken，within the servo，to minimize this difference[5].

The last basic component of a CNC system is the machine tool. It accomplishes the processing steps to transform the starting workpiece into a completed part.

Words and Expressions

simultaneous	[ˌsɪml'teɪnɪəs]	adj.	同时发生的
servo	['sɜːvəʊ]	n.	伺服
in-process			过程中的
gauging	['geɪdʒɪŋ]	n.	测量
unmanned	[ʌn'mænd]	adj.	无人操纵的,无人的
transducer	[trænz'djuːsə]	n.	转换器,传感器
submission	[səb'mɪʃn]	n.	服从
host	[həʊst]	n.	主机
convert	[kən'vɜːt]	v.	使转变

proprietary	[prə'praɪətri]	n.	所有权(所有人)
		adj.	专利的(所有的)
architecture	['ɑːkɪtektʃə]	n.	体系结构
reliability	[rɪˌlaɪə'bɪlɪti]	n.	可靠性
servo-amplifier			伺服增幅器
discrete	[dɪ'skriːt]	adj.	不连续的,离散的

NOTES

[1] Current CNC systems allow simultaneous servo position and velocity control of all axis, monitoring of controller and machine tool performance, online programming with graphical assistance, in-process cutting process monitoring, and in-process part gauging for completely unmanned machining operations.

当前的 CNC 系统可以同时对所有的轴进行伺服位置和速度控制,监测控制器和机床性能,借助图形辅助功能完成在线编程,监测切削过程,并能对完全无人操控的切削中的在制零件进行测量。

[2] Additional instructions are usually included, such as spindle speed, feed rate, cutting tool selection, and other functions.

其他的指令也通常包含在内,诸如主轴速度、进给率、切削刀具的选择,以及其他功能。

[3] The machine control unit(MCU) consists of a microcomputer and related control hardware that stores the program of instructions and executes it by converting each command into mechanical actions of machine tool, one command at a time.

机床控制单元(MCU)包括一个微型计算机及相关控制设备,用来存储和执行程序指令,即将每一个指令转换成机床的机械动作,一个指令一次。

[4] The numerical commands are converted into signal voltage by the MCU unit and sent to servo-amplifiers, which process and amplify them to the high voltage levels required by the drive motors.

MCU 单元将数字命令转换成信号电压,传送到伺服放大器,通过处理并放大成驱动电机所需要的更高的电压。

[5] The difference between the required position and the actual position is detected by comparison circuit and the action is taken, within the servo, to minimize this difference.

比较电路检测实际位置和所需位置之差并执行动作,减少这一伺服系统的误差。

第 9 单元　计算机数字控制(CNC)系统

当前的 CNC 系统可以同时对所有的轴进行伺服位置和速度控制,监测控制器和机床性能,借助图形辅助功能完成在线编程,监测切削过程,并能对完全无人操控的切削中的在制零件进行测量,其中大部分功能都可由制造商提供选择。

典型的 CNC 机床有 5 个基本单元:(1) 输入介质;(2) 机床控制单元;(3) 伺服驱动单元;(4) 反馈传感器;(5) 机械的机床单元。5 个组成部分之间的总体关系如图 9-1 所示。

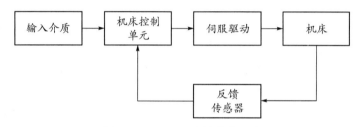

图 9-1　CNC 机床的基本组成

输入介质包含控制程序,也叫作零件程序,是指导机床运行的、详细的、按部就班的指令。单个指令指的是切削刀具相对于固定工件的工作台的位置,其他的指令也通常包含在内,如主轴速度、进给率、切削刀具的选择,以及其他功能。

为了方便向机床控制单元提交程序,其编码要在合适的媒介上进行,早期的 NC 系统,最常用的输入介质是打孔卡片。目前,在现代化的车间里,打孔卡片已经基本上被新的存储技术所代替,更先进的方法是与计算机直接连接,这种方法叫作直接数字控制(DNC)。有了 DNC 技术,一组 NC 或 CNC 机床就可以由一个主机同时控制。

CNC 系统的核心部件就是控制,机床控制单元(MCU)包括一个微型计算机及相关控制设备,用来存储和执行程序指令,即将每一个指令转换成机床的机械动作,一个指令一次。

专用控制单元具有闭环结构,系统由制造商定制,常常包含严密的监测电路、算法和控制程序。这一类型的控制造价昂贵,但是可靠性非常高。近几年来,有向所谓的开环结构控制发展的趋势——也就是说,由普通零部件和软件构成的控制。对一些控制系统的制造者来说,这意味着基于个人计算机(PC 机)的控制实现了。

在伺服驱动单元中,机床的驱动分为主轴和进给驱动机构。主轴和进给驱动电机以及它们的伺服放大器是伺服驱动单元的组成部分。MCU 单元处理数据并产生离散的数字位置指令传送给进给驱动,产生速度指令传送给主轴驱动。MCU 单元将数字命令转换成信号电压,传送到伺服放大器,通过处理并放大成驱动电机所需要的更高的电压。

反馈系统的功能是为控制提供运动控制系统状况的信息,当驱动机构移动时,传感

器对它们的实际位置进行测量。比较电路检测实际位置和所需位置之差,并执行动作,以减少这一伺服系统误差。

CNC 系统的最后一个基本组成部分是机床,它完成初始工件到成品的加工过程。

READING MATERIAL

Safety and Maintenance for CNC Machine

Safety is always a major concern in a metal-cutting operation. CNC equipment is automated and very fast, and consequently it is a source of hazards. The hazards have to be located and the personnel must be aware of them in order to prevent injuries and damage to the equipment. Main potential hazards include: rotating parts, such as the spindle, the tool in the spindle, chuck, part in the chuck, and the turret with the tools and rotating clamping devices; movable parts, such as the machining center table, lathe slides, tailstock, and tool carousel; errors in the program such as improper use of the G00 code in conjunction with the wrong coordinate value, which can generate an unexpected rapid motion; an error in setting or changing the offset value, which can result in a collision of the tool with the part or the machine; and a hazardous action of the machine caused by unqualified changes in a proven program[1]. To minimize or avoid hazards, try the following preventive actions:

（1）Keep all of the original covers on the machines as supplied by the machine tool builder.

（2）Wear safety glasses, gloves, and proper clothing and shoes.

（3）Do not attempt to run the machine before you are familiar with its control.

（4）Before running the program, make sure that the part is clamped properly.

（5）When proving a program, follow these safety procedures:

① Run the program using the machine "Lock" function to check the program for errors in syntax and geometry.

② Slow down rapid motions using the RAPID OVERRIDE switch or dry run the program.

③ Use a single-block execution to confirm each line in the program before executing it.

④ When the tool is cutting, slow down the feed rate using the FEED OVERRIDE switch to prevent excessive cutting conditions.

（6）Do not handle chips by hand and do not use chip hooks to break long curled chips. Program different cutting conditions for better chip control. Stop the machine if you need to proper clean the chips.

(7) If there is any doubt that the insert will break under the programmed cutting conditions, choose a thicker insert or reduce feed or depth of cut[2].

(8) Keep tool overhang as short as possible, since it can be a source of vibration that can break the insert.

(9) When supporting a large part by the center, make sure that the hole-center is large enough to adequately support and hold the part.

(10) Stop the machine when changing the tools, indexing inserts, or removing chips.

(11) Replace dull or broken tools or inserts.

(12) Write a list of offsets for active tools, and clear(set to zero) the offsets for tools removed from the machine[3].

(13) Do not make changes in the program if your supervisor has prohibited your doing so.

(14) If you have any safety-related concerns, notify your instructor or supervisor immediately.

Normally, CNC machines are very safe to use as they are designed to be as safe as possible. One of the main advantages of CNC machines is that they are much safer than manually operated machines.

Words and Expressions

chuck	[tʃʌk]	n.	卡盘,吸盘
turret	[ˈtʌrɪt]	n.	转台,转动架,转塔刀架
clamp	[klæmp]	vt.	夹住,固定
slide	[slaɪd]	n.	滑块,滑座,钻杆导槽
tailstock	[ˈteɪlstɒk]	n.	尾架
carousel	[ˌkærəˈsel]	n.	轮盘,转盘
in conjunction with			连同,与……协力
offset	[ˈɔːfset]	n.	偏移
collision	[kəˈlɪʒən]	n.	碰撞,冲突
preventive	[prɪˈventɪv]	adj.	预防的
syntax	[ˈsɪntæks]	n.	句法
insert	[ɪnˈsɜːt]	n.	插入物,刀片
		v.	插入
vibration	[vaɪˈbreɪʃn]	n.	振动

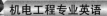

NOTES

[1] Main potential hazards include: rotating parts, such as the spindle, the tool in the spindle, chuck, part in the chuck, and the turret with the tools and rotating clamping devices; movable parts, such as the machining center table, lathe slides, tailstock, and tool carousel; errors in the program such as improper use of the G00 code in conjunction with the wrong coordinate value, which can generate an unexpected rapid motion; an error in setting or changing the offset value, which can result in a collision of the tool with the part or the machine; and a hazardous action of the machine caused by unqualified changes in a proven program.

主要潜在的危险包括：旋转部件，如主轴、主轴内刀具、卡盘、卡盘内工件、带刀具的转塔刀架以及旋转的夹具装置；运动部件，如加工中心的工作台、车床滑板、尾架以及刀具轮盘；程序错误，如 G00 代码的不正确使用和错误的坐标值，会产生意想不到的快速移动；设置或更改偏移值出错，可能会导致刀具与工件或刀具与机床之间的碰撞；随意更改已验证的程序，也会导致机床产生危险的动作。

[2] If there is any doubt that the insert will break under the programmed cutting conditions, choose a thicker insert or reduce feed or depth of cut.

如果觉得刀片在已编程的切削状态下有可能折断的话，可选择更厚的刀片以及减少进给或切削的深度。

[3] Write a list of offsets for active tools, and clear(set to zero) the offsets for tools removed from the machine.

列出所使用刀具的偏移量清单，从机床取下刀具，清除（设置为 0）其偏移量。

阅读材料

计算机数控(CNC)机床的安全与维护

在金属切削操作中，安全性一直特受关注。由于计算机数控设备自动化程度高且速度很快，所以它是一个危险源。为了防止人员受伤和对设备造成损害，必须找出危险的根源所在，而且操作人员必须提高警惕。主要潜在的危险包括：旋转部件，如主轴、主轴内刀具、卡盘、卡盘内工件、带刀具的转塔刀架以及旋转的夹具装置；运动部件，如加工中心的工作台、车床滑板、尾架以及刀具轮盘；程序错误，如 G00 代码的不正确使用和错误的坐标值，会产生意想不到的快速移动；设置或更改偏移值出错，可能会导致刀具与工件或刀具与机床之间的碰撞；随意更改已验证的程序，也会引起机床产生危险动作。为了减少或避免危险，应尽量遵循以下保护措施：

1. 使用机床制造商提供的机床原厂的防护罩。

2.佩戴安全眼镜、手套,并穿着合适的衣服和鞋子。

3.在不熟悉机床控制之前不要开动机床。

4.在程序运行之前确保零件已正确夹紧。

5.验证一个程序时,遵循以下安全步骤:

(1) 启用机床锁定功能运行程序,检查程序中的语法错误和几何轨迹;

(2) 使用 RAPID OVERRIDE 快速倍率开关降低速度或空运行程序;

(3)采用单程序段执行来确认程序中的每一行;

(4)刀具切削时,用 FEED OVERRIDE 进给倍率开关来减慢进给率,防止超负荷切削。

6.禁止用手处理切屑,不要用切屑钩子弄断长而卷曲的切屑。编制不同的切削状态程序,以便更好地控制切屑。如果要彻底清除切屑,应关闭机床。

7.如果觉得刀片在已编程的切削状态下有可能折断的话,可选择更厚的刀片以及减少进给或切削的深度。

8.尽可能保持刀具悬出短些,因为它可能成为导致刀片折断的振动根源。

9.当用顶尖支撑大零件时,要确保中心孔足够大,以支撑和夹紧零件。

10.换刀、查找刀片或清理切屑时要关闭机床。

11.更换已磨损或毁坏的刀具和刀片。

12.列出所使用刀具的偏移量清单,从机床取下刀具,清除(设置为0)其偏移量。

13.在未得到主管许可情况下不得擅自更改程序。

14.如果发现有安全方面的问题,应立即通知技术指导或主管。

通常情况下,使用 CNC 机床是非常安全的,因为在设计时就已经尽量考虑到了它的安全性。CNC 机床的主要优点之一就是它比手动操作的机床要安全得多。

Unit 10

Robot

扫码可见本项目
☞ 相关参考资料

The industrial robot is a tool that is used in the manufacturing environment to increase productivity. It can perform jobs that might be hazardous to the human worker. One of the first industrial robots was used to replace the nuclear fuel rods in nuclear power plants. The industrial robot can also operate on the assembly line such as placing electronic components on a printed circuit board. Thus, the human worker can be relieved of the routine operation of this tedious task. Robots can also be programmed to defuse bombs, to serve the handicapped, and to perform functions in numerous applications in our society.

A robot is a reprogrammable, multifunctional manipulator designed to move parts, materials, tools, or special devices through variable preprogrammed locations for the performance of a variety of different tasks.

Preprogrammed locations are paths that the robot must follow to accomplish work. At some of these locations, the robot will stop and perform some operation, such as assembly of parts, spray painting, or welding. These preprogrammed locations are stored in the robot's memory and are recalled later for continuous operation. Furthermore, these preprogrammed locations, as well as other program data, can be changed later as the work requirements change. Thus, with regard to this programming feature, an industrial robot is like a computer very much.

The robotic system can also control the work cell of the operating robot. The work cell of the robot is the total environment in which the robot must perform its task. Included within this cell may be the robot manipulator, controler, a work table, safety features, or a conveyor. In addition, signals from outside devices can communicate with the robot.

The manipulator, which does the physical work of the robotic system, consists of two sections: the mechanical section and the attached appendage. The manipulator also has a base to which the appendages are attached. The base of the manipulator is usually fixed to the floor of the work area. Sometimes, though, the base may be movable. In this case, the base is attached to either a rail or a track, allowing the manipulator to be moved from one location to another.

The appendage is the arm of the robot. It can be either a straight, movable arm or a jointed arm and gives the manipulator its various axes of motion. The jointed arm is also known as an articulated arm. At the end of the arm, a wrist is connected. The wrist is made up of additional axes and a wrist flange. The wrist flange allows the robot user to connect different tooling to the wrist for different jobs. The manipulator's axes allow it to perform work within a certain area. This area is called the work cell of the robot, and its size corresponds to the size of the manipulator. As the robot's physical size increases, the size of the work cell must also increase.

The movement of the manipulator is controlled by actuators, or drive systems. They allow the various axes to move within the work cell. The drive system can use electric, hydraulic, or pneumatic power. The energy developed by the drive system is converted to mechanical power by various mechanical drive systems. The drive systems are coupled through mechanical linkages. These linkages, in turn, drive the different axes of the robot. The mechanical linkages may be composed of chains, gears, and ball screws.

The controller is used to control the robot manipulator's movements as well as to control peripheral components within the work cell. The user can program the movements of the manipulator into the controller through the use of a hand-held teach pendant. This information is stored in the memory of the controller for later recall.

The controller is also required to communicate with peripheral equipment within the work cell. For example, a controller has an input line. When the machine cycle is completed, the input line turns on, telling the controller to position the manipulator so that it can pick up the finished part. Then, a new part is picked up by the manipulator and placed into the machine. Next, the controller signals the machine to start operation.

The controller can send electric signals over communication lines. This two-way communication between the robot manipulator and the controller maintains a constant update of the location and operation of the system. The controller also has the job of communicating with the different plant computers. The communication link establishes the robot as part of a computer-assisted manufacturing (CAM) system. The microprocessor-based systems operate in conjunction with solid-state memory devices. These memory devices may be magnetic bubbles, random-access memory, floppy disks, or magnetic tape.

The power supply is the unit that supplies power to the controller and the manipulator. Two types of power are delivered to the robotic system. One type of power is the AC power for operation of the controller. The other type of power is used for driving the various axes of the manipulator. For example, if the robot manipulator is controlled by hydraulic or pneumatic drives, control signals are sent to these devices, causing motion of the robot.

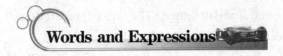

Words and Expressions

hazardous	[ˈhæzədəs]	adj.	危险的,冒险的
tedious	[ˈtiːdɪəs]	adj.	沉闷的,单调乏味的
defuse	[diːˈfjuːz]	vt.	拆除(爆炸物)信管,使缓和
handicapped	[ˈhændɪkæpt]	n.	残疾人,身体有缺陷的人
appendage	[əˈpendɪdʒ]	n.	附加物,附属物
axes	[ˈæksɪz]	n.	轴
articulated	[ɑːˈtɪkjuleɪtɪd]	adj.	铰接的,枢接的,有关节的
flange	[flændʒ]	n.	边缘,轮缘,凸缘,法兰
pendant	[ˈpendənt]	n.	垂环;垂饰;下垂物
peripheral	[pəˈrɪfərəl]	adj.	外围的
state-of-the-art		adj.	达到最新技术发展水平的
conjunction	[kənˈdʒʌŋkʃn]	n.	连接词,连合,关联
hydraulic	[haɪˈdrɒlɪk]	adj.	水力的,水压的
pneumatic	[njuːˈmætɪk]	adj.	气动的(有空气的,气体学的)

第 10 单元　机器人

工业机器人是一种提高制造业生产力的工具。它可以承担那些对人类可能有危险的工作。最早的工业机器人就曾用来在核能发电厂中更换核燃料棒。工业机器人也能在装配线上工作,如安装印刷电路板上的电子元器件。这样,人们就可以从这种单调的工作中解脱出来。机器人还能拆除炸弹,为伤残人士服务,为我们的社会做各种各样的工作。

机器人是一个可以重复编程的、多功能的机械手,可以在各个预编程位置移动零件、材料、工具或其他特殊装置,完成各种不同的工作。

预编程位置是指机器人完成工作时必须遵循的路径。在某些预编程位置,机器人会停下来进行一些操作,例如安装零件、喷漆或焊接。这些预编程位置存储在机器人的存储器中,以便随时调出进行连续的操作。如果工作要求改变了,这些预编程位置连同其他的编程数据也能随之改变。这些编程特征使得工业机器人与计算机非常类似。

机器人系统可以控制工作机器人的工作单元。机器人的工作单元是机器人执行任务时的工作环境。工作单元包括机器人的机械手、控制器、工作台、安全装置以及传动装置。此外,机器人应该能与外界信号进行交流。

机器人的机械手完成机器人系统的具体工作,它包括两个部分:机械部分和附属部分。附属部分安装在机械手的基座上。基座固定在工作现场的地板上。但有时基座也是可以移动的,在这种情况下,基座安放在轨道上,用于把机械手从一个位置移到另外的位置。

附属部分是机器人的手臂。它可能是一个直的可以移动的手臂,也可能是一个铰接的手臂,为机械手提供多根工作轴。铰接的手臂也就是有关节连接的手臂。手臂的端部连有一个手腕。手腕装在另一根轴上并装有法兰盘。在法兰盘上可以连接不同的工具,完成不同的工作。机械手上的轴允许机械手在一个特定的区域里工作。这个区域叫作机器人的工作单元,它的大小取决于机械手的大小。如果机器人的尺寸增加,工作单元的尺寸也会增加。

机械手的运动由驱动器或驱动系统控制。它们驱动各轴在工作单元内旋转。驱动系统可以是电力的、液压的,也可以是气动的。驱动系统产生的动力经各种不同的机构转换成机械能。各种驱动系统经机械传动装置相连。这些由链条、齿轮和滚珠丝杠组成的机械传动装置驱动机器人的各轴。

控制器用于控制机械手的运动和工作单元内的外部设备。可以通过悬挂的手持键盘把机械手的运动程序输入控制器。这些数据存储在控制器的内存中,以备将来调用。

控制器还要能与工作单元内的外部设备进行通信。例如,控制器有一条输入线路。加工完成时输入线路接通,告诉控制器让机械手在指定的位置挑出加工好的零件。机械手把一个新的零件送入机器后,控制器发出信号,重新开始加工。

控制器可以在通信线路上传送电信号。这种机械手和控制器之间的双向通信不断地更新系统的位置和操作。控制器的工作还包括与不同设备的计算机进行通信。这种通信连接使机器人成为计算机辅助制造系统的一部分。微处理器系统使用固态存储装置。这些存储装置可能是磁泡、随机存储器、软盘及磁带。

控制器和机械手的动力由动力源供给。机器人系统一般使用两种动力:一种是提供给控制器的交流电;另一种动力源用于驱动机械手的各轴。例如,如果机械手是由液压或气动驱动控制的,这些装置将接收控制信号,使机器人产生运动。

READING MATERIAL

Welding Robot

The welding robot is an industrial robot engaged in welding. Industrial robot is a versatile, reproducible manipulator with three or more programmable axes for industrial automation. In order to adapt itself to different uses, the mechanical interface of the robot's last axis is usually a connecting flange, which can be attached to different tools or end effectors. The welding robot is an industrial robot of which a welding tong or torch is in the end, so that it can weld, cut or hot spray.

With the development of electronic technology, computer technology, numerical control and robot technology, since the beginning of the 1960s, automatic welding robot began to be used for production; its technology has become increasingly mature. It has mainly the following advantages:

(1) to stabilize and improve the quality of welding, which can be reflected in the form of value;

(2) to improve labor productivity;

(3) to improve the labor intensity of workers who can work in a harmful environment;

(4) to reduce requirements for workers' operating techniques;

(5) to shorten the product replacement of the preparation cycle, which reduce the corresponding equipment investment.

Therefore, welding robot has been widely used.

Welding robot mainly includes two parts robot and welding equipment. The robot consists of the robot body and the control cabinet (hardware and software). Take arc welding as an example, the welding equipment consists of the welding power (including its control system), wire feeder, welding gun and other components. For intelligent robot it should also have a sensing system, such as laser or camera sensors and their control devices.

In addition, if the workpiece in the entire welding process is without displacement, you can use the fixture to locate the workpiece on the table; this system is the most simple. However, in the actual production, many workpiece in the welding need to change position, so that the weld in a better position (posture) is welded. For this situation, the positioning machine and the robot can separately move, that is the robot welds again after the displacement machine changing position; and they can also be moving at the same time, the displacement machine is changing position while the robot is welding, that is often said that the displacement machine and robot coordinate to move. At this time the movement of the positioning machine and the movement of the robot are compound movement, so that the movement of the welding gun relative to the workpiece can meet the weld trajectory, and can meet the requirements of the welding speed and the welding gun's posture. In fact, the shaft of the positioner has become part of the robot, and this welding robot system can be up to 7~20 axes or more.

Words and Expressions

torch	[tɔːtʃ]	n.	火把;火炬
stabilize	['steɪbəlaɪz]	vt.	使稳固,使安定
		vi.	稳定,安定

displacement	[dɪsˈpleɪsmənt]	n.	取代;移位;位移
component	[kəmˈpəʊnənt]	n.	部件;组件;成分
device	[dɪˈvaɪs]	n.	装置;策略;图案
fixture	[ˈfɪkstʃə(r)]	n.	设备;固定装置
workpiece	[ˈwɜːkpiːs]	n.	工件;轧件;工件壁厚
posture	[ˈpɒstʃə(r)]	n.	姿势;态度
		vi.	摆姿势
separately	[ˈseprətli]	adv.	分别地;分离地;个别地
coordinate	[kəʊˈɔːdɪneɪt]	adj.	并列的;同等的
		vt.	调整;整合
shaft	[ʃɑːft]	n.	[机]轴;箭杆
		vt.	利用
end effector			末端执行器
welding tong			焊钳
hot spray			热喷雾
electronic technology			电子技术
numerical control			数控
labor productivity			劳动生产率
labor intensity			劳动强度
preparation cycle			准备周期
welding equipment			焊接设备
arc welding			电弧焊
wire feeder			送丝机;送丝装置
welding gun			焊枪
intelligent robot			智能机器人
sensing system			传感系统
camera sensors			相机传感器
positioning machine			定位机
displacement machine			变位机
compound movement			复合运动
weld trajectory			焊接轨迹

参考译文

焊接机器人

　　焊接机器人是从事焊接的工业机器人。工业机器人是一种多用途的、可重复编程的自动控制操作机,具有三个或更多可编程的轴,用于工业自动化领域。为了适应不同的用途,机器人最后一个轴的机械接口通常是一个连接法兰,可接装不同的工具或末端执行器。焊接机器人在工业机器人的末端法兰上装接焊钳或焊(割)枪,使之能进行焊接、切割或热喷涂。

随着电子技术、计算机技术、数控及机器人技术的发展,自动焊接机器人从 19 世纪 60 年代开始用于生产以来,其技术已日趋成熟,主要有以下优点:

(1) 稳定和提高焊接质量,能将焊接质量以数值的形式反映出来;

(2) 提高劳动生产率;

(3) 改善工人劳动强度,可在有害环境下工作;

(4) 降低了对工人操作技能的要求;

(5) 缩短了产品改型换代的准备周期,减少了相应的设备投资。

因此,焊接机器人在各行各业已得到了广泛的应用。

焊接机器人主要包括机器人和焊接设备两部分。机器人由机器人本体和控制柜(硬件及软件)组成。而焊接装备,以弧焊为例,则由焊接电源(包括其控制系统)、送丝机、焊枪和其他部分组成。智能机器人还应有传感系统,如激光或相机传感器及其控制装置等。

另外,如果工件在整个焊接过程中无须变位,就可以用夹具把工件定位在工作台面上,这种系统是最简单的。但在实际生产中,多数的工件在焊接时需要变位,使焊缝处在较好的位置(姿态)下焊接。对于这种情况,定位机与机器人可以分别运动,即变位机变位后机器人再焊接;也可以是同时运动,即变位机变位的同时机器人进行焊接,也就是常说的变位机与机器人协调运动。这时定位机的运动及机器人的运动复合,使焊枪相对于工件的运动既能满足焊缝轨迹又能满足焊接速度及焊枪姿态的要求。实际上,这时定位机的轴已成为机器人的组成部分,这种焊接机器人系统可以多达 7~20 个轴或更多。

Unit 11

Mechatronics

扫码可见本项目
☞ 参考资料

The word mechatronics was first introduced by the senior engineer of a Japanese company, Yaskawa, in 1969, as a combination of "mecha" of mechanisms and "tronics" of electronics, and the company was granted trademark rights on the word in 1971. The word soon received broad acceptance in industry and, in order to allow its free use, Yaskawa decided to abandon his rights on the word in 1982. The word has taken a wider meaning since then, and is now widely used as a technical jargon word to describe a philosophical idea in engineering technology, more than technology itself. For this wider concept of mechatronics, a number of definitions has been proposed in scientific literature, differing in the particular characteristics, which each definition is intended to emphasize. The most commonly used one emphasizes synergy: Mechatronics is synergistic integration of mechanical engineering, electronics and intelligent computer control in design and manufacture of products and processes.

The development of mechatronics has gone through three stages. The first stage corresponds to the years when this term was introduced. During this stage, technologies used in mechatronic systems developed rather independently and individually. With the beginning of the eighties, a synergistic integration of different technologies started taking place, the notable example is optoelectronics (i.e. an integration of optics and electronics). The concept of hardware/software co-design also started in those years. The third and the last stage can also be considered as the beginning of the mechatronics age since early nineties. The most notable aspect of the third stage is the increased use of computational intelligence in mechatronic products and systems. It is due to this development that we can now talk about Machine Intelligence Quotient(MIQ). Another important achievement of the third stage is the possibility of miniaturization of components; in the form of micro actuators and micro sensors(i.e. micro mechatronics).

As it shown in Figure 11 - 1, the computer disk drive is one of the best examples of mechatronic design because it exhibits quick response, precision, and robustness. According to mechatronic principles clothes washer features a sensor-based feedback

Figure 11 - 1　Examples of mechatronic systems

control that maintains correct water temperature no matter the load size.

Computer control device carries out the following basic functions:

1. Control of mechanical movement process in mechatronic module or in multi-measured system on-line with current sensor data analysis.

2. Arrangements to control Mechatronics System(MS) functional movements that is to coordinate simultaneously MS mechanical movements and external processes. Discrete input/output devices are always used for external process control.

3. Interaction with operator via human-machine interface off-line and on-line interaction at the moment of MS movement.

4. Data exchange between peripheral devices, sensors and other devices of the system.

The main task of MS is to transform input information from the top level of control into purposeful mechanical movement with its control based on principle of feedback[2]. It is notable that electric power(less often hydraulic or pneumatic) is used as intermediate power form in modern systems.

The core of mechatronic approach consists in integrating of two or more components probably of different physical nature into a uniform functional module. There are basic advantages of mechatronic approach in comparison with traditional means of automation:

• Rather low cost owing to a high degree of integration, unification and standardization of all elements and interfaces;

• Ability to perform complicated and precise movements(of high quality) owing to application intellectual control methods;

• High reliability, durability and noise immunity;

• Constructive compactness of modules (up to miniaturization in micro machines);

• Improved overall dimension and dynamic characteristics of machines owing to simplification of kinematics' circuits;

• Opportunity to rebuild functional modules to sophisticated systems and complexes according to specific purposes of the customer.

Words and Expressions

mechatronics	[meka'trɒnɪks]	*n.*	机电一体化;机械电子学
jargon	['dʒɑːgən]	*n.*	行话,术语
synergistic	[ˌsɪnə'dʒɪstɪk]	*adj.*	协同的;协作的,协同作用的
quotient	['kwəʊʃ(ə)nt]	*n.*	[数]商;系数;份额
sophisticated	[sə'fɪstɪkeɪtɪd]	*adj.*	复杂的;精致的;富有经验的
kinematics	[ˌkɪnɪ'mætɪks;ˌkaɪn-]	*n.*	运动学;动力学
senior	['siːnɪə]	*adj.*	高级的;年长的;年资较深的
grant	[grɑːnt]	*vt.*	授予;允许;承认
		vi.	同意
trademark	['treɪdmɑːk]	*n.*	商标,标志
philosophical	[fɪlə'sɒfɪk(ə)l]	*adj.*	哲学的(等于 philosophic)
notable	['nəʊtəbl]	*adj.*	值得注意的,显著的;著名的
optoelectronics	[ˌɒptəʊɪlek'trɒnɪks]	*n.*	光电子
miniaturization	[ˌmɪnɪətʃəraɪ'zeɪʃn]	*n.*	小型化,微型化
actuator	['æktjʊeɪtə]	*n.*	[自]促动器
robustness	[rəʊ'bʌstnɪs]	*n.*	[计]稳健性
unification	[ˌjuːnɪfɪ'keɪʃn]	*n.*	统一;一致;联合

NOTES

[1] The word has taken a wider meaning since then, and is now widely used as a technical jargon word to describe a philosophical idea in engineering technology, more than technology itself.

自此,这个术语有了更广泛的含义,同时被广泛用来描述工程技术中的哲学思想,而不是技术本身。

[2] The main task of MS is to transform input information from the top level of control into purposeful mechanical movement with its control based on principle of feedback.

机电一体化系统的主要任务是将输入信息从控制端转换为一定的机械运动,并根据反馈原理进行控制。

第11单元 机电一体化

"机电一体化"最初是由一家日本公司的高级工程师 Yaskawa 在 1969 年提出。作为机械系统的"机"和电子工业的"电子学"的结合体,它于 1971 年被安川公司申请授予商标权,并在工业界很快得到了广泛的认可。为了使其能够在工业中得到自由使用,1982 年安川公司决定放弃对该术语的商标权。自此,它有了更广泛的含义,现在被广泛用作技术术语来描述工程技术中的哲学思想,而不是技术本身。对于机电一体化的概念,不同的科学文献给出了不同的定义,每一个定义都有不同的特点,该特点在定义中得到强化。最常用的一种定义强调了协同的概念:机电一体化是机械工程、电子和智能计算机控制在产品和工艺设计、制造中的协同集成。

机电一体化的发展经历了三个阶段。第一阶段对应于提出该术语的年份。在这一阶段,机电一体化系统中的技术发展相对独立。八十年代初,不同技术的协同集成开始出现,如光电子学(即光学与电子的集成)就是一个不可忽视的例子。硬件/软件协同设计的概念也始于此。第三个阶段也可以认为是 90 年代初,也是机电一体化时代的开始,这个阶段最显著的特点是在机电产品和系统中增加了计算机智能化。正是由于第三阶段的发展,才引入了机器智商(MIQ)的概念。第三阶段的另一个重要成就是使得组件小型化成为可能,如微型执行器和微型传感器(即微型机电一体化)。

图11-1 机电一体化系统示例

如图 11-1 所示,计算机磁盘驱动器是机电一体化设计的最佳示例之一,因为它具有快速响应、精确性和坚固性特点。根据机电一体化原理,洗衣机充分利用传感器的反馈控制,可以确保无论衣服多还是少,都能调节保持合适的水况。

计算机控制装置执行以下基本功能:

1.通过电流传感器数据分析,在线控制机电模块或测量系统的机械运动过程。

2.控制机电一体化系统功能的实施,即同时协调系统机械运动和外部过程。采用离散输入/输出装置控制外部过程。

3.通过人机离线和在线界面与操作人员进行配合响应。

4.系统外部设备、传感器和其他设备之间的数据交换。

机电一体化系统的主要任务是将输入信息从控制端转换为有目的的机械运动,并根据反馈原理进行控制。值得注意的是,在现代系统中,系统中的动力源较多使用电力,而较少使用液压或气动。

机电一体化方法的核心在于将两个或多个物理性质不同的组件集成到一个统一的功能模块中。与传统的自动化方法相比,机电一体化方法有其基本优势:

● 由于所有元素和接口的高度集成、统一和标准化,成本相当低;

- 由于应用了智能控制方法,能够高质量地执行复杂而精确的运动;
- 高可靠性、耐用性和抗噪声性;
- 模块结构紧凑(微型机可以达到最小型化);
- 由于简化了运动单元,改善了机器的外形尺寸和动态特性;
- 可以根据客户的特定目的,再次将功能模块重新构建成其他复杂系统。

READING MATERIAL

Modern Trends of Mechatronic Systems

World production of MS is constantly increasing and expands new spheres. Today mechatronic modules and systems find wide application in the following areas:

- Machine-tool construction and equipment for automation of technological processes;
- Robotics(industrial and special);
- Aviation, space and military techniques;
- Motor car construction [for example, antilocking brake system (ABS), systems of car movement stabilization and automatic parking];
- Non-conventional vehicles(electro bicycles, cargo carriages, electro scooters, invalid carriages);
- Office equipment(for example, copy and fax machines);
- Computer facilities(for example, printers, plotters, disk drives);
- Medical equipment(rehabilitation, clinical, service);
- Home appliances(washing, sewing and other machines);
- Micro machines(for medicine, biotechnology, means of telecommunications);
- Control and measuring devices and machines;
- Photo and video equipment;
- Simulators for training of pilots and operators;
- Show-industry(sound and illumination systems).

Impetuous development of mechatronics as new scientific and technical direction in the nineties was caused by a lot of factors among which there are the following key factors: trends of global industrial development; development of fundamental basic and mechatronic methodology (the base scientific ideas, essentially new technical and technological decisions); activity of experts in research and educational spheres[1].

It is possible to distinguish the following tendencies to change and key requirements of the world market in the considered area:

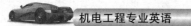

- Necessity for production and service of equipment according to the international system of the quality standards stated in the Standard ISO 9000;
- Internationalization of scientific and technical production market and, as a consequence;
- Necessity for active introduction of forms and methods of international engineering and putting new technologies into practice;
- Increasing role of small and average industrial enterprises in economy owing to their ability to quick and flexible reaction to changing requirements of the market;
- Rapid development of computer systems and technologies, telecommunications. Intellectualization of mechanical movement control systems and technological functions of modern machines appear as a consequence of this common tendency.

The analysis of these specified tendencies shows that it is impossible to achieve a new level of the basic process equipment following traditional approaches. Development of mechatronics as interdisciplinary scientific and technical sphere besides obvious technique-technological difficulties also has a lot of new managerial and economic problems[2]. Modern enterprises starting to develop and produce mechatronic products should solve the following fundamental problems:

- Structural integration of mechanical, electronic and information departments (which as a rule, work independently) into a uniform creative staff;
- Education and training of engineers specialized in mechatronics and managers able to organize integration and supervise work of strictly specialized experts with different qualifications;
- Integration of information technologies from various scientific and technical fields(mechanic, electronics, computer control) into a uniform toolkit to provide computer support of mechatronic problems;
- Standardization and unification of all used elements and processes at designing and manufacturing mechatronic systems.

Solution of the listed problems frequently demands that we should put an end to traditions, which were established in management before, that the average managers, who got used to solve only particular problems, should overcome their ambitions. For this reason, average and small enterprises, which have flexible structure, turned out to be more prepared to start manufacturing mechatronic production.

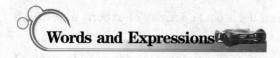
Words and Expressions

aviation   [eɪvɪ'eɪʃn]   *n.*   航空;飞行术;飞机制造业

scooters	[ˈskuːtəs]	n.	踏板车（scooter 的复数）
rehabilitation	[ˌriːəˌbɪlɪˈteɪʃn]	n.	康复
clinical	[ˈklɪnɪkl]	adj.	临床的；诊所的
biotechnology	[ˌbaɪə(ʊ)tekˈnɒlədʒi]	n.	［生物］生物技术
illumination	[ɪˌljuːmɪˈneɪʃn]	n.	照明
impetuous	[ɪmˈpetʃuəs]	adj.	冲动的；鲁莽的；猛烈的
methodology	[ˌmeθəˈdɒlədʒi]	n.	方法学，方法论
intellectualization	[ˌɪntəlektjuəlaɪˈzeɪʃn]	n.	智能化
interdisciplinary	[ˌɪntədɪsəˈplɪnəri]	adj.	各学科间的；跨学科的

NOTES

[1] Impetuous development of mechatronics as new scientific and technical direction in the nineties was caused by a lot of factors among which there are the following key factors: trends of global industrial development; development of fundamental basic and mechatronic methodology (the base scientific ideas, essentially new technical and technological decisions); activity of experts in research and educational spheres.

90 年代机电一体化作为新的科学技术得到迅速发展，其中主要有以下几个关键影响因素：全球工业发展趋势；机电一体化基础方法论的发展（基础科学思想，本质上是新技术应用和技术决策）；研究和教育领域专家的作用。

[2] Development of mechatronics as interdisciplinary scientific and technical sphere besides obvious technique-technological difficulties also has a lot of new managerial and economic problems.

机电一体化作为一个跨学科的科学技术领域，除了技术难度明显外，还存在着许多新的管理和经济学问题。

阅读材料

机电系统的现代化趋势

随着全球机电一体化产品数量不断增加，并被应用到越来越多的新领域。今天机电一体化的模块和系统主要在以下领域得到了广泛的应用：

- 用于工艺过程自动化的机床设备；
- 机器人（包含工业和特殊用途）；
- 航空、航天和军事技术；
- 汽车［如防抱死制动系统（ABS）、汽车运动稳定系统和自动停车系统］；

- 非传统车辆(含电动自行车、货车、电动踏板车、病人用车辆);
- 办公设备(如复印机和传真机);
- 计算机设施(如打印机、绘图仪、磁盘驱动器等);
- 医疗设备(康复、临床、服务等);
- 家用电器(洗衣机、缝纫机和其他电器等);
- 微型机(用于医学、生物技术、电信);
- 控制和测量装置;
- 照片和视频设备;
- 飞行员和操作员模拟培训器;
- 演艺业(音响和照明系统)。

90 年代机电一体化作为新的科学技术得到迅速发展,其中主要有以下几个关键影响因素:全球工业发展趋势;机电一体化基础方法论的发展(基础科学思想,本质上是新技术应用和技术决策);研究和教育领域专家的作用。

在应用领域内,我们可以根据如下变化趋势来满足国际化市场的关键要求:

- 很有必要根据 ISO 9 000 标准规定的国际质量标准体系生产和服务设备;
- 科学技术市场的国际化;
- 很有必要积极引进国际化工程并实施新技术;
- 基于小型工业企业能够快速灵活地应对市场变化,进一步提升其在经济发展中的作用;
- 促进计算机系统技术、电信技术的快速共同发展。机械控制系统的智能化和新一代机器的技术性能正体现了这种共同发展趋势。

对上述特定趋势的分析表明,按照传统的方法,不可能达到基础工艺设备的新水平。机电一体化作为一个跨学科的科学技术领域,除了技术难度明显外,还存在着许多新的管理和经济学问题。现代企业在开发和生产机电一体化产品时,必须解决以下基本问题:

- 将机械、电子和信息部门(通常独立工作)整合为同一部门,便于统筹创新发展;
- 对专门从事机电一体化的工程师,以及能够组织整合和督促不同专业人员工作的管理人员进行教育和培训;
- 将来自各科技领域(机械、电子、计算机控制)的信息技术集成到统一的信息工具库中,为机电一体化问题提供计算机支持;
- 在设计和制造过程中,对机电系统中所用的元件和工艺进行标准化和统一化。

要想解决如上所列举的问题,往往要求管理人员打破以往固有的管理模式,更需要那些习惯于只解决特定问题的一般管理人员克服他们的困难。正因为如此,普通和小型企业的机构设置灵活,对于生产机电一体化产品更感兴趣。

Unit 12

An Overview of PLC

扫码可见本项目
参考资料

Initially industries used relays to control the manufacturing processes. The relay control panels had to be regularly replaced, consumed lot of power and it was difficult to figure out the problems associated with it. To sort these issues, programmable logic controller(PLC) was introduced.

Figure 12 – 1　Graphical representation of PLC

What is PLC?

As shown in Figure 12 – 1, PLC is a digital computer used for the automation of various electro-mechanical processes in industries. These controllers are specially designed to survive in harsh situations and shielded from heat, cold, dust, and moisture etc.[1]. PLC consists of a microprocessor which is programmed using the computer language.

The program is written on a computer and is downloaded to the PLC via cable. These loaded programs are stored in non-volatile memory of the PLC. During the transition of relay control panels to PLC, the hard-wired relay logic was exchanged for the program fed by the user. A visual programming language known as the Ladder Logic was created to program the PLC.

PLC Hardware

The hardware components of a PLC system are CPU, Memory, Input/Output, Power supply unit, and programming device.

● **CPU**—Keeps checking the PLC controller to avoid errors. They perform functions including logic operations, arithmetic operations, computer interface and many more.

● **Memory**—System (ROM) stores the data permanently for the operating system. RAM stores the information of the status of input and output devices, and the values of timers, counters and other internal devices.

● **I/O section**—Input keeps a track on field devices which includes sensors, switches.

● **O/P Section**—Output has a control over the other devices which includes

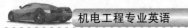

motors，pumps，lights and solenoids.

- **Power supply**—Certain PLCs have an isolated power supply. But，most of the PLCs work at 220VAC or 24VDC.

- **Programming device**—This device is used to feed the program into the memory of the processor. The program is first fed to the programming device and later it is transmitted to the PLC's memory.

- **System buses**—Buses are the paths through which the digital signal flows internally of the PLC. The four system buses are as follows：

  ➤ Data bus is used by the CPU to transfer data among different elements.

  ➤ Control bus transfers signals related to the action that are controlled internally.

  ➤ Address bus sends the location's addresses to access the data.

  ➤ System bus helps the I/O port and I/O unit to communicate with each other.

Working of PLC

The programmable logic controller functions in four steps.

- **Input scan**：The state of the input is scanned which is connected externally. The inputs include switches，pushbuttons，and proximity sensors，limit switches，pressure switches.

- **Program scan**：The loaded program is executed to carry out the function appropriately.

- **Output scan**：The input sources have a control over the output ports to energize or deenergize them. The outputs include solenoids，valves，motors，actuator，and pumps. Depending on the model of PLC，these relays can be transistors，triacs or relays.

- **Housekeeping**

PLC Applications

The simple suitable application is a conveyor system. The requirements of the conveyor systems are as follows：

- A programmable logic controller is used to start and stop the motors of the conveyor belt.

- The conveyor system has three segmented conveyor belts. Each segment is run by a motor.

- To detect the position of a plate，a proximity switch is positioned at the segment's end.

- The first conveyor segment is turned ON always.

- The proximity switch in the first segment detects the plate to turn ON the second conveyor segment.

- The third conveyor segment is turned ON when the proximity switch detects the plate at the second conveyor.

● As the plate comes out of the detection range，the second conveyor is stopped after 20 secs.

● When the proximity switch fails to detect the plate，the third conveyor is stopped after 20 secs.

relay	[ˈriːleɪ]	*n.*	继电器;中继设备
harsh	[hɑːʃ]	*adj.*	严酷的;粗糙的
volatile	[ˈvɒlətaɪl]	*adj.*	不稳定的
arithmetic	[əˈrɪθmətɪk]	*n.*	算术,算法
solenoid	[ˈsəʊlənɔɪd]	*n.*	[电]螺线管;螺线形电导管
energize	[ˈenədʒaɪz]	*vt.*	激励;使活跃;供给……能量
		vi.	活动;用力
triac	[ˈtraɪæk]	*n.*	三端双向可控硅元件
conveyor	[kənˈveɪə]	*n.*	输送机,[机]传送机;传送带
proximity	[prɒkˈsɪməti]	*n.*	接近

These controllers are specially designed to survive in harsh situations and shielded from heat，cold，dust，and moisture etc.

这些控制器经过专门设计可以在恶劣的环境中正常工作,并能屏蔽热、冷、灰尘和湿气等。

第 12 单元　PLC 的概述

最初,工业界使用继电器来控制制造过程,然而继电器控制面板必须定期更换、耗电量大,且在使用过程中出现问题很难找出。为了解决上述这些问题,研究者们引入了可编程逻辑控制器(PLC)的概念。

什么是 PLC?

如图 12-1 所示,PLC 是一种用于工业机电自动化的数字计算机。这些控制器经过专门设计可以在恶劣的环境中正常工作,并能屏蔽热、冷、灰尘和湿气等。PLC由使用计算机语言编程的微处理器组成。

图 12-1　PLC 的典型外观图

在计算机上编写程序,并通过电缆下载到 PLC 上。这些加载的程序存储在 PLC 的非易失性存储器中。在从继电器控制面板传送到 PLC 的过程中,硬接线继电器的逻辑与用户提供的程序进行交换,这样就形成了一种被称为梯形逻辑的可视化编程语言用来对 PLC 进行编程。

PLC 硬件

PLC 系统的硬件包括 CPU、存储器、输入/输出、电源单元和编程设备。

● CPU——持续检查 PLC 控制器以避免出错。它执行的功能包括逻辑运算、算术运算、计算机接口等。

● 存储器——系统只读存储器(ROM)永久存储操作系统数据。系统随机存储器(RAM)存储输入和输出设备的状态信息,计时器、计数器和其他内部设备的值。

● I/O 部分——输入保持跟踪现场设备,包括传感器、开关。

● O/P 部分——输出控制其他设备,包括电机、泵、灯和电磁阀。

● 电源——某些 PLC 具有独立的电源。但是,大多数 PLC 工作电源为 220 伏交流电或 24 伏直流电。

● 编程设备——用于将程序输入处理器的内存。程序首先被输入编程设备,然后被传送到 PLC 的存储器。

● 系统总线——总线是数字信号在 PLC 内部流动的路径。四条系统总线是:

➤ 数据总线用于 CPU 在不同元件间进行数据传输。

➤ 控制总线传输与内部控制动作相关的信号。

➤ 地址总线发送位置地址以访问数据。

➤ 系统总线帮助 I/O 端口和 I/O 单元相互通信。

PLC 的工作

PLC 的工作主要包括如下四步:

● 输入扫描:扫描外部连接的输入状态。输入包括开关、按钮和接近传感器、限位开关、压力开关。

● 程序扫描:加载程序以执行相应的功能。

● 输出扫描:输入源对输出端口进行控制,使其通电或断电。输出包括电磁阀、阀门、电机、执行器和泵。根据 PLC 的型号,这些继电器可以是晶体管、三端双向可控硅元件或继电器。

● 管理指令

PLC 应用

PLC 最简单的应用是输送系统。输送系统的要求如下:

● PLC 用于启动和停止输送带的电机。

● 输送系统有三个分段输送带,每一段由一台电机控制。

● 为了检测板的位置,在分段输送带的末端放置一个接近开关。

● 第一段输送带始终打开。

● 当第一段输送带的接近开关检测到板时,第二段输送带打开。

● 当接近开关检测到第二个输送带上的板时,第三段输送带打开。

- 当板超出检测范围时,第二段输送带在 20 秒后停止。
- 当接近开关未能检测到板时,第三段输送带在 20 秒后停止。

READING MATERIAL

Advantages & Disadvantages of PLCs

PLCs are well adapted to a range of automation tasks, as shown in Figure 12 - 2, PLC can be in a control panel. These are typically industrial processes in manufacturing where the cost of developing and maintaining the automation system is high relative to the total cost of the automation, and where changes to the system would be expected during its operational life. PLCs contain input and output devices compatible with industrial pilot

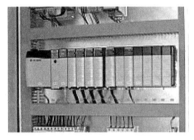

Figure 12 - 2　PLC installed in a control panel

devices and controls; little electrical design is required, and the design problem centers on expressing the desired sequence of operations[1].

PLC applications are typically highly customized systems, so the cost of a packaged PLC is low compared to the cost of a specific custom-built controller design. On the other hand, in the case of mass-produced goods, customized control systems are economical. This is due to the lower cost of the components, which can be optimally chosen instead of a "generic" solution, and where the non-recurring engineering charges are spread over thousands or millions of units[2].

For high volume or very simple fixed automation tasks, different techniques are used. For example, a cheap consumer dishwasher would be controlled by an electromechanical cam timer costing only a few dollars in production quantities.

From above,we can conclude advantages & disadvantages of PLCs:

Advantages

- PLCs can be programmed easily which can be understood clearly well.
- They are fabricated to survive vibrations, noise, humidity, and temperature.
- The controller has the input and output for interfacing.
- High dependability and strong anti-jamming ability.
- Simply programming and convenient usage.
- Complete functions and strong commonality.
- Simple design and installation as well as convenient maintenance.

Disadvantages

- It is a tedious job when replacing or bringing any changes to it.

- Skillful work force is required to find its errors.
- Lot of effort is put to connect the wires.
- The hold up time is usually indefinite when any problem arises.

Words and Expressions

customize	[ˈkʌstəmaɪz]	vt.	定做,按客户具体要求制造
optimal	[ˈɒptɪml]	adj.	最佳的;最理想的
generic	[dʒəˈnerɪk]	adj.	类的;一般的
recur	[rɪˈkɜː]	vi.	复发;重现;采用;再来;循环;递归
dependability	[dɪˌpendəˈbɪlɪti]	n.	可靠性;可信任
commonality	[kɒməˈnælɪti]	n.	公共;共性
tedious	[ˈtiːdiəs]	adj.	沉闷的;冗长乏味的
indefinite	[ɪnˈdefɪnət]	adj.	不确定的;无限的;模糊的

NOTES

[1] PLCs contain input and output devices compatible with industrial pilot devices and controls, little electrical design is required, and the design problem centers on expressing the desired sequence of operations.

PLC 系统包含与工业试验装置和控制装置兼容的输入和输出装置,几乎不需要进行电气设计,设计问题集中在表达所需的操作顺序上。

[2] This is due to the lower cost of the components, which can be optimally chosen instead of a "generic" solution, and where the non-recurring engineering charges are spread over thousands or millions of units.

这是由于部件成本较低,可以通过优化选择而不是"通用"的解决方案,并且非常规工程费用可以被分摊到数千或者数百万个单元里面。

阅读材料

PLC 的优缺点

PLC 系统能够很好地适应不同的自动化任务,如图 12-2 所示,它可以安装在控制面板中。在典型的制造工业中,自动化系统的开发和维护成本相对于自动化的总成本很高,并且还会随着预期的系统运行寿命的要求变化而变化。PLC 系统包含与工业试验装置和控制装置兼容的输入和输出装置,几乎不需要进行电气设计,设计问题集中在表达所需的操作顺序上。

　　PLC 应用程序通常是对用户高度通用的,因此与特定定制控制器设计成本相比,PLC 的开发成本较低。通常只有在批量生产的条件下,特定定制的控制系统才会比较经济,这是由于部件成本较低,可以通过优化选择而不是仅采用"通用"的解决方案,非常规的工程费用还可以被分摊到数千甚至数百万个单元里面。

图 12 - 2　安装在控制面板中的 PLC

　　对于大容量或非常简单单一的自动化任务,可以使用不同的自动化技术。例如,一台便宜的家用洗碗机采用一台机电凸轮定时器控制,其生产成本仅为几美元。

　　综上,我们可总结出 PLC 的优缺点如下:

优点:

● PLC 易于理解和编程。

● PLC 的设计可承受振动、噪音、湿度和温度。

● 控制器具有接口的输入和输出。

● 可靠性高,抗干扰能力强。

● 编程简单,使用方便。

● 功能齐全,通用性强。

● 设计简单,安装、维护方便。

缺点:

● 在对 PLC 更换或对其进行更改时,比较枯燥乏味。

● 发现运行中的错误需要大量的经验。

● 连接电线需要付出大量的工作。

● 运行中出现问题时,延期时间通常不确定。

Unit 13

Introduction to Modern Control System

扫码可见本项目
参考资料

This contribution of modern control system is dedicated to present the main approaches and their mathematical support of development the control systems theory. The underlying idea is that, at present, Control Theory is an interdisciplinary area of research where many mathematical concepts and methods work together to produce an impressive body of important applied mathematics[1]. Control systems theory witnessed different stages and approaches.

The word control has two main meanings. First, it is understood as the activity of testing or checking that a physical device has a satisfactory behavior. Secondly, to control is to act, to implement decisions that guarantee that device behaves as desired. Although the mathematical formulation of control problems, based on the mathematical models of physical systems, is intrinsically complex, the fundamental ideas in control theory are enough simple and very intuitive. These key ideas can be found in nature, in the evolution and behavior of living beings. There are three fundamental concepts in control theory. The first one is that of feedback. The second key concept in control theory is that of the need for fluctuations. The third very important concept in control theory is that of optimization.

Several factors provided the stimulus for the development of modern control theory:

a. The necessary of dealing with more realistic models of system.

b. The shift in emphasis towards optimal control and optimal system design.

c. The continuing developments in digital computer technology.

d. The shortcoming of previous approaches.

e. Recognition of the applicability of well-known methods in other fields of knowledge.

The transition from simple approximate models, which are easy to work with, to more realistic models, produces two effects. First, a large number of variables must be included in the models. Second, a more realistic model is more likely to contain nonlinearities and time-varying parameters. Previously ignored aspects of the system, such as interactions with feedback through the environment, are more likely to be included.

With an advancing technological society, there is a trend towards more ambitious goals. This also means dealing with complex system with a large number of interacting components. The need for greater accuracy and efficiency has changed the emphasis on control system performance. The classical specifications in terms of percent overshoot, setting time, bandwidth, etc. have in many cases given way to optimal criteria such as minimum energy, minimum cost, and minimum time operation. Optimization of these criteria makes it even more difficult to avoid dealing with unpleasant nonlinearities. Optimal control theory often dictates that nonlinear time-varying control laws are used, even if the basic system is linear and time-invariant.

The continuing advances in computer technology have had three principal effects on the controls field. One of these relates to the gigantic supercomputers.

The second impact of the computer technology has to do with the proliferation and wide availability of the microcomputers in homes and the work place, classical control theory was dominated by graphical methods because at the time that was the only way to solve certain problems. Now every control designer has easy access to powerful computer packages for systems analysis and design. The old graphical methods have not yet disappeared, but have been automated. They survive because of the insight and intuition that they can provide, some different techniques are often better suited to a computer. Although a computer can be used to carry out the classical transform-inverse transform methods, it is used usually more efficient for a computer to integrate differential equations directly.

The third major impact of the computers is that they are now so commonly used as just another component in the control systems. This means that the discrete-time and digital system control now deserves much more attention than it did in the past[2].

Modern control theory is well suited to the above trends because its time-domain techniques and its mathematical language(matrices, linear vector spaces, etc.) are ideal when dealing with a computer. Computers are a major reason for the existence of state variable methods.

Modern control theory is a recent development in the field of control. Therefore, the name is justified at least as a descriptive title. However, the foundations of modern control theory are to be found in other well-established fields. The field of linear algebra contributes heavily to modern control theory. This is due to the concise notation, the generality of the results, and the economy of thought that linear algebra provides.

Words and Expressions

dedicate	[ˈdedɪkeɪt]	vt.	致力;献身
underlying	[ˌʌndəˈlaɪɪŋ]	adj.	潜在的;根本的;在下面的;优先的
witness	[ˈwɪtnəs]	vt.	目击;证明;为……作证
intrinsical	[ɪnˈtrɪnsɪkəl]	adj.	本质的;固有的;内在的
fluctuation	[ˌflʌktʃuˈeɪʃn]	n.	起伏,波动
ambitious	[æmˈbɪʃəs]	adj.	野心勃勃的;有雄心的;热望的;炫耀的
overshoot	[ˌəʊvəˈʃuːt]	n.	超越目标
bandwidth	[ˈbændwɪdθ]	n.	[电子][物]带宽;[通信]频带宽度
nonlinearity	[ˌnɒnlɪnɪˈærɪti]	n.	[数]非线性;非线性特征
time-invariant		adj.	不变时的
matrice	[ˈmeɪtrɪs]	n.	[数]矩阵
linear algebra		n.	[数]线性代数
notation	[nəʊˈteɪʃn]	n.	符号,记号

NOTES

[1] The underlying idea is that, at present, Control Theory is an interdisciplinary area of research where many mathematical concepts and methods work together to produce an impressive body of important applied mathematics.

其基本思想在于阐明目前控制理论是一个跨学科的研究领域,涉及多种数学概念和数学方法,从而产生了重要的令人印象深刻的应用数学。

[2] This means that the discrete-time and digital system control now deserves much more attention than it did in the past.

这就意味着离散时间和数字式的控制系统现在比过去更受到关注。

第13单元 现代控制系统简介

现代控制系统致力于展示控制系统理论发展的主要方法及其数学基础。其基本思想在于阐明目前控制理论是一个跨学科的研究领域,涉及多种数学概念和数学方法,从而产生了重要的令人印象深刻的应用数学。控制系统理论经历了不同的阶段和方法。

单词control有两个主要含义。首先,它被用来测试或检查系统设备是否具有令人满意的性能。第二,控制就是行动,用以保证设备按预期性能运行。虽然基于物理系统数学模型控制问题的数学公式本质上很复杂,但控制理论的基本思想足够简单,非常直观。这些关键思想可以在自然界、生物的进化和行为中发现。控制理论有三个基本概念:第一是反馈;第二个关键概念是波动;第三个非常重要的概念是优化。

下列几方面为促进现代控制理论发展的主要因素:

1. 处理更加真实的系统模型的必要性;

2. 强调最佳控制和最佳系统设计的升级;

3. 数字化计算机技术的持续发展;

4. 当前技术的不成熟;

5. 认可将熟知的方法在其他知识领域的应用。

从易解决的简单近似模型到真实模型的转变引发了两种效果:首先,模型必须包括很多的变量;其次,更为真实的模型往往具有非线性和时变参数。以前常常忽略的一些系统问题,例如关联问题以及通过环境形成的反馈等,现在却需要考虑。

在现代科技高度发达的社会中,朝着更加宏伟目标发展趋势是很明显的,这也意味着要处理具有大量相互关联部件的复杂系统。高精确度与高效率的需要改变了控制系统的性能要求重点。百分超调量、调节时间、带宽等方面的典型技术指标,已经在很多情况下让位于最优化准则,如最小能耗、最低成本、最短时间操作等。依据这些准则的最优化使得想要避免处理令人不愉快的非线性事件变得更加困难。即使基础受控系统是线性的和不随时间变化的,最优控制理论也经常应用非线性、随时间变化的控制规律。

不断发展的计算机技术在控制领域产生了三个最重要的影响,其中一项是与超级计算机相关。

计算机技术的第二个影响与微型计算机在家庭和工作场所的普及和广泛使用相关。经典控制理论中图解方法占主导地位,这是因为当时图解法是解决某些问题的唯一方法。现在,每一个控制工程设计人员都很容易获得功能强大的计算机软件包,用于进行系统的分析和设计工作。传统的图解法并没有消失,已经可以进行自动分析,它们之所以能存在在于所具有的直观性和指导性。然而,一些完全不同的技术通常更适合于计算机,比如计算机可用于执行经典的变换—反变换运算,然而用计算机对微分方程直接进行积分更加有效。

计算机技术的第三个影响,是如今计算机的应用如此普遍,就像是控制系统中的其他常规元件一样。这就意味着离散时间和数字式的控制系统现在比过去更应该受到关注。

现代控制理论特别适应上述的发展潮流。这是因为时间领域技术和数学表达语言(矩阵、线性向量空间等)在计算机上的应用是非常理想的。计算机的发展也是导致状态变量法产生的一个主要原因。

现代控制理论是控制领域的新发展。因此,可以说它是名副其实的。然而,现代控制理论的基础却应该在其他一些发展成熟的领域中寻找。线性代数对于现代控制理论的发展功不可没,这归功于线性代数所提供的简明的符号、通用的结果和高效的思路。

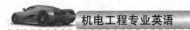

READING MATERIAL

The Application of Modern Control System

There are wide applications of modern control system on mechanic industries.

Mechanism of Surface Finish Production

There are basically five mechanisms which contribute to the production of a surface which have been machined. There are:

(1) The basic geometry of the cutting process. In, for example, single point turning the tool will advance a constant distance axially per revolution of the work piece and the resultant surface will have on it, when viewed perpendicularly to the direction of tool feed motion, a series of cusps which will have a basic form which replicates the shape of the tool in cut.

(2) The efficiency of the cutting operation. It has already been mentioned that cutting with unstable built-up-edges will produce a surface which contains hard built-up-edge fragments which will result in a degradation of the surface finish[1]. It can also be demonstrated that cutting under adverse conditions such as apply when using large feeds small rake angles and low cutting speeds, besides producing conditions which continuous shear occurring in the shear zone, tearing takes place, discontinuous chips of uneven thickness are produced, and the resultant surface is poor. This situation is particularly noticeable when machining very ductile materials such as copper and aluminum.

(3) The stability of the machine tool. Under some combinations of cutting conditions: workpiece size, method of clamping, and cutting tool rigidity relative to the machine tool structure, instability can be set up in the tool which causes it to vibrate. Under some conditions the vibration will be built up and unless cutting is stopped, considerable damage to both the cutting tool and workpiece may occur. This phenomenon is known as chatter.

(4) The effectiveness of removing sword. In discontinuous chip production machining, such as milling or turning of brittle materials, it is expected that the chip(sword) will leave the cutting zone either under gravity or with the assistance of a jet of cutting fluid and that they will not influence the cut surface in any way. However, when continuous chip production is evident, unless steps are taken to control the swarf it is likely that it will impinge on the cut surface and mark it. Inevitably, this marking beside a looking unattractive, often results in a poorer surface finishing.

(5) The effective clearance angle on the cutting tool. For certain geometries of

minor cutting edge relieve the clearance angles it is possible to cut on the major cutting edge and burnish on the minor cutting edge. This can produce a good surface finish but, of course, it is strictly a combination of metal cutting and metal forming and is not to be recommended as a practical cutting method.

Surface Finishing and Dimensional Control

Products that have been completed to their proper shape and size frequently require some type of surface finishing to enable them to satisfactorily fulfill their function. In some cases, it is necessary to improve the physical properties of the surface material for resistance to penetration or abrasion. In many manufacturing processes, the product surface is left with dirt, chips, grease, or other harmful materials upon it. Assemblies that are made of different materials, or from the same materials processed in different manners, many require some special surface treatment to provide uniformity of appearance.

Surface finishing many sometimes become an intermediate step processing. For instance, cleaning and polishing are usually essential before any kind of plating process. Some of the cleaning procedures are also used for improving surface smoothness on mating parts and for removing burrs and sharp corners, which might be harmful in later use. Another important need for surface finishing is for corrosion protection in a variety of environments. The type of protection procedure will depend largely upon the anticipated exposure, with due consideration to the material being protected and the economic factors involved.

Satisfying the above objectives necessitates the use of main surface-finishing methods that involve chemical change of the surface mechanical work affecting surface properties, cleaning by a variety of methods, and the application of protective coatings, organic and metallic.

In the early days of engineering, the mating of parts was achieved by machining one part as nearly as possible to the required size, machining the mating part nearly to size, and then completing its machining, continually offering the other part to it, until the desired relationship was obtained. If it was inconvenient to offer one part to the other part during machining, the final work was done at the bench by a fitter, who scraped the mating parts until the desired fit was obtained, the fitter therefore being a "fitter" in the literal sense. It is obvious that the two parts would have to remain together, and in the event of one having to be replaced, the fitting would have to be done all over again. In these days, we expect to be able to purchase a replacement for a broken part, and for it to function correctly without the need for scraping and other fitting operations.

When one part can be used "off the shelf" to replace another of the same dimension and material specification, the parts are said to be interchangeable[2].

Words and Expressions

axial	[ˈæksiəl]	adj.	轴的;轴向的
resultant	[rɪˈzʌltənt]	adj.	由此导致的,因而发生的
perpendicularly	[ˌpɜːpənˈdɪkjuləli]	adv.	垂直地;直立地
cusp	[kʌsp]	n.	尖头;尖端
replicate	[ˈreplɪkeɪt]	vt.	复制;折叠
degradation	[ˌdegrəˈdeɪʃn]	n.	退化;降格,降级
transient	[ˈtrænziənt]	adj.	短暂的;路过的
		n.	瞬变现象;过往旅客;候鸟
sword	[sɔːd]	n.	刀
swarf	[swɔːf]	n.	木屑;金属的切屑
impinge	[ɪmˈpɪndʒ]	v.	对……有明显作用(或影响)
penetration	[peniˈtreɪʃn]	n.	渗透;突破;侵入;洞察力
uniformity	[juːnɪˈfɔːməti]	n.	均匀性;一致;同样
scrap	[skræp]	n.	碎片;残余物;少量

NOTES

[1] It has already been mentioned that cutting with unstable built-up-edges will produce a surface which contains hard built-up-edge fragments which will result in a degradation of the surface finish.

　　如用不稳定的切削瘤切削将会加工出包含坚硬的切削瘤碎片的表面,而这些将会直接导致表面粗糙度等级降低。

[2] When one part can be used "off the shelf" to replace another of the same dimension and material specification,the parts are said to be interchangeable

　　如果一个部件能被用作备用件去替换另一个同样尺寸和材料特性的部件,我们就称之为它们具有互换性。

■ 阅读材料

现代控制系统的应用

现代控制系统在机械工业中有广泛的应用。

表面粗糙度的加工技术

在机加工表面,有五种影响其表面粗糙度的技术:

1. 切削过程的基础几何学。如,在单点车削时,工件每转一周,刀具就沿轴线方向进给一个固定的距离。从垂直刀具进给的方向观察,所得到的表面上有很多尖角,这些尖角的形状与切削刀具的形状相同。

2. 切削操作的效率。如用不稳定的切削瘤切削将会加工出包含坚硬的切削瘤碎片的表面,而这些将会直接导致粗糙度等级降低。由试验证明,在采用进给量大、前角小、切削速度低的不利情况下,除了产生不稳定的切削瘤外,切削过程也会不稳定。同时,在切削过程中进行的也不再是切削,而是撕裂,导致厚度不均匀,产生不连续的切削,加工出表面质量差。在切削加工延展性良好的金属材料(如铜和铝时)时,这种情况就更为突出。

3. 加工刀具的稳定性。在许多联合切削的情况下,工件的大小、夹紧的方法和切削刀具相对于机床结构的坚硬度,不稳定性是建立在使其变化的工具基础上的。在某些情况下,这种变化将达到并保持很长一段时间,在另外一些情况下,除非切削停止,这种变化将会同时对切削刀具和工件产生破坏,这种现象就是有名的刀振。

4. 刀刃的移动效率。在不连续的产品加工过程中,如易碎材料的磨削或旋转,我们期望碎片在重力作用或在冷却液的喷射作用下离开切削区域,这样就不会影响切削表面。然而,在连续切削中会产生明显的碎屑,这时就需逐步控制刀刃,否则很有可能影响切削表面并在其上留下记号。不可避免地,这些不美观的记号时常导致较差的表面粗糙度。

5. 切削刀具的有效清除角。由于副切削刃的几何特征减轻了清除角,使得在主切削面上主切削刃切削和副切削刃打磨变得可能。这样能加工出良好的表面粗糙度,但是,严格意义上,这是一种金属切削和金属成型的综合技术,而不仅仅是一种切削方法。

表面精整加工与尺寸控制

产品在被加工成所需的外形和尺寸时,经常需要采用各种表面精整加工方法,使其能够达到比较令人满意的性能。在有些情况下,很有必要通过提高材料表面的物理特性来抵抗腐蚀和磨损。在许多制造过程中,产品表面都残留有污垢、碎屑、油渍以及其他有害的材料。假设上述残留是由不同种金属材料,或是由同一种金属材料在不同的加工方式中所造成的,大多数情况下我们都需要一些特殊的表面处理技术来获得均匀的外表面。

有时表面精整加工只是中间阶段处理,例如,在电镀前必须进行清洁和磨光工序。有些清洁程序是为了改善配合处表面的光滑程度,或是清除对后续工序会产生有害作用的毛刺和尖角。表面精整加工的另一重要用途就是为了在多种环境下防腐蚀。这种保护很大程度上依赖于预期的工作环境情况,同时要考虑材料需要被保护的程度和其所包含的经济因素。

为了使材料的表面达到令人满意的效果,必然要使用表面精整技术,包括改变材料工作表面化学特性;各种方法清洗;应用有机保护膜和金属保护膜。

在早期工程中,零部件的装配是这样完成的:加工一个部件使其尽可能地达到所要求的尺寸,然后继续加工另一部件,直到获得所要求的配合关系。如果在加工一个部件时不方便同时提供另一部件,那么最后的工作将由装配工完成,他需要刮削装配部件直到获得所要求的配合关系。因此,装配工也就完全体现了"装配工"字面上的意思。很显然,这两部件将被强迫配合在一起,重要的是当其中一个部件需要更换时,装配不得不全部重新完成。此时,我们希望能够在不需要进行刮磨和其他装配作业的前提下,就能方便地更换一个已经坏掉的零件,并且能保持原有功能。

如果一个部件能被用作备用件去替换另一个同样尺寸和材料特性的部件,那么我们就说它们具有互换性。

Unit 14

Pressure Sensor Technology

扫码可见本项目
参考资料

Sensor is an electronic device or object capable of detecting real-life conditions and interpreting what it has detected into data that the device can understand[1].

A good example of a sensor is a motion sensor that may be connected to a computer to help detect motion within a building. Other sensors may detect heat, light, radiation, and sound.

Figure 14 - 1 Digital air pressure sensor

Figure 14 - 2 Miniature digital barometric pressure sensor

A pressure sensor is a device for pressure measurement of gases or liquids. As shown in Figures 14 - 1 and 14 - 2, pressure is an expression of the force required to stop a fluid from expanding, and is usually stated in terms of force per unit area. A pressure sensor usually acts as a transducer; it generates a signal as a function of the pressure imposed. For the purposes of this article, such a signal is electrical.

Pressure sensors are used for control and monitoring in thousands of everyday applications. Pressure sensors can also be used to indirectly measure other variables such as fluid/gas flow, speed, water level, and altitude. Pressure sensors can alternatively be called pressure transducers, pressure transmitters, pressure senders, pressure indicators, piezometers and manometers, among other names.

Pressure sensors can vary drastically in technology, design, performance, application suitability and cost. A conservative estimate would be that there may be over 50 technologies and at least 300 companies making pressure sensors worldwide.

There is also a category of pressure sensors that are designed to measure in a dynamic mode for capturing very high speed changes in pressure[2]. Example applications for this type of sensor would be in the measuring of combustion pressure in an engine cylinder or in a gas turbine. These sensors are commonly manufactured out of piezoelectric materials such as quartz.

Some pressure sensors are pressure switches, which turn on or off at a particular pressure, as shown in Figure 14-3. For example, a water pump can be controlled by a pressure switch so that it starts when water is released from the system, reducing the pressure in a reservoir.

Figure 14-3　Silicon piezoresistive pressure sensors

Pressure sensors can be classified in terms of pressure ranges they measure, temperature ranges of operation, and most importantly the type of pressure they measure. Pressure sensors are variously named according to their purpose, but the same technology may be used under different names.

● **Absolute pressure sensor**

This sensor measures the pressure relative to perfect vacuum.

● **Gauge pressure sensor**

This sensor measures the pressure relative to atmospheric pressure. A tire pressure gauge is an example of gauge pressure measurement; when it indicates zero, then the pressure it is measuring is the same as the ambient pressure.

● **Vacuum pressure sensor**

This term can cause confusion. It may be used to describe a sensor that measures pressures below atmospheric pressure, showing the difference between that low pressure and atmospheric pressure, but it may also be used to describe a sensor that measures absolute pressure relative to a vacuum.

● **Differential pressure sensor**

This sensor measures the difference between two pressures, one connected to each side of the sensor. Differential pressure sensors are used to measure many properties, such as pressure drops across oil filters or air filters, fluid levels(by comparing the pressure above and below the liquid) or flow rates(by measuring the change in pressure across a restriction). Technically speaking, most pressure sensors are really differential pressure sensors; for example, a gauge pressure sensor is merely a differential pressure sensor in which one side is open to the ambient atmosphere.

● **Sealed pressure sensor**

This sensor is similar to a gauge pressure sensor except that it measures pressure relative to some fixed pressure rather than the ambient atmospheric pressure(which varies according to the location and the weather).

There are many applications for pressure sensors: pressure sensing, altitude sensing, flow sensing, level/depth sensing, leak testing.

Words and Expressions

miniature	[ˈmɪnətʃə]	adj.	微型的,小规模的
barometric	[ˌbærəʊˈmetrɪk]	adj.	气压的
impose	[ɪmˈpəʊz]	vi.	施加影响
transducer	[trænzˈdjuːsə]	n.	[自]传感器,[电子]变换器,[电子]换能器
piezometer	[ˌpaɪˈzɒmɪtə]	n.	[物]压强计
manometer	[məˈnɒmɪtə]	n.	压力计;[流]测压计;血压计
conservative	[kənˈsɜːvətɪv]	adj.	保守的
piezoelectric	[piːˌeɪzəʊˈlektrɪk]	adj.	[电]压电的
silicon	[ˈsɪlɪkən]	n.	[化学]硅;硅元素
vacuum	[ˈvækjʊəm]	n.	真空;空间;真空吸尘器
restriction	[rɪˈstrɪkʃn]	n.	限制;约束;束缚

NOTES

[1] Sensor is an electronic device or object capable of detecting real-life conditions and interpreting what it has detected into data that the device can understand.

从字面意义上理解,传感器就是能够检测实际情况并将其检测到的信息转变为设备能够接收的数据信息的电子设备或物体。

[2] There is also a category of pressure sensors that are designed to measure in a dynamic mode for capturing very high speed changes in pressure.

还有一种压力传感器,可以在动态模式下测量压力的高速变化。

第14单元　压力传感器技术

从字面意义上理解,传感器就是能够检测实际情况并将其检测到的信息转变为设备能够接收的数据信息的电子设备或物体。

典型的传感器如运动传感器,它可以连接到计算机,协助检测建筑物内的运动。还有其他一些传感器可以检测热、光、辐射和声音信息。

图14-1　数字气压传感器

图14-2　微型数字气压传感器

压力传感器是测量气体或液体压力的装置。如图14-1、14-2所示,压力是指阻止流体膨胀所需的力,通常用单位面积上受到的力来表示。压力传感器通常用作转换器,它产生一定的信号,该信号为所施加的压力。在本文中,这种信号是电信号。

日常生活中,成千上万的压力传感器被用于控制和监测。压力传感器也可用于间接测量其他变量,如流体/气体流量、速度、水位和高度。压力传感器也可以称为压力转换器、压力变送器、压力传送器、压力指示器、压强计和压力计等。

压力传感器在技术、设计、性能、应用适用性和成本等方面有很多变化。保守估计,全球有50多种压力传感器技术和至少300家压力传感器制造公司。

图14-3　硅压阻式压力传感器

还有一种压力传感器,可以在动态模式下测量压力的高速变化。这种类型传感器的典型应用是测量发动机气缸或燃气轮机中的燃烧压力。这些传感器通常由石英等压电材料制成。

还有一些压力传感器仅作为压力开关,可以在特定压力下打开或关闭,如图14-3所示。例如,水泵可以由压力开关控制,这样当压力开关打开时,水就从水库中释放出去,从而降低水库中的压力。

压力传感器可以根据其测量的压力

范围、工作温度范围以及压力类型进行分类。压力传感器根据其用途有不同的名称,但相同的技术也可用于不同的名称。

- **绝对压力传感器**

该传感器测量相对于理想真空的压力。

- **表压传感器**

该传感器测量相对于大气压的压力。轮胎压力表是表压传感器的一个典型应用,当它显示数据为零时,它测量的压力与环境压力相同。

- **真空压力传感器**

这个词容易引起混淆。该传感器可用于测量低于大气压的压力,它显示该低压和大气压之间的差异,但它也可用于测量相对于真空的绝对压力。

- **压差传感器**

此类传感器测量两个压力之间的差异,两个压力源分别连接到传感器的两侧。压差传感器用于测量多种性能,例如机油滤清器或空气滤清器的压降、液位(通过比较液体上方和下方的压力)或流量(通过测量受限制后的压力变化)。从技术层面上来说,大多数压力传感器实际上都是压差传感器,如表压传感器就是一个一侧与环境大气相通的压差传感器。

- **密封压力传感器**

此类传感器类似于表压传感器,只是它测量的压力是相对于某个固定压力,而不是相对于环境大气压力的(大气压力随环境位置和天气变化而变化)。

压力传感器在如下场合有着广泛的应用:压力传感、高度传感、流量传感、液位/深度传感、泄漏测试等。

READING MATERIAL

Electric Motors

The first electric motors were simple electrostatic devices described in experiments by Scottish monk Andrew Gordon and American experimenter Benjamin Franklin in the 1740s. The theoretical principle behind them, Coulomb's law, was discovered but not published, by Henry Cavendish in 1771. This law was discovered independently by Charles-Augustin de Coulomb in 1785, who published it so that it is now known with his name. The invention of the electrochemical battery by Alessandro Volta in 1799 made possible the production of persistent electric currents. After the discovery of the interaction between such a current and a magnetic field, namely the electromagnetic interaction by Hans Christian in 1820 much progress was soon made. It only took a few weeks for Andre Marie Ampere to develop the first formulation of the electromagnetic interaction and present the Ampere's force law, that described the production of mechanical force by the interaction of an electric current and a magnetic field[1]. The first demonstration of the effect with a rotary motion was given by Michael Faraday in 1821. A free-hanging wire was dipped into a pool of mercury, on which a permanent magnet (PM) was placed. When a current was passed through the wire, the wire rotated around the magnet, showing that the current gave rise to a close circular magnetic field around the wire. This motor is often demonstrated in physics experiments, substituting brine for(toxic) mercury.

In 1827, Hungarian physicist Anyos Jedlik started experimenting with electromagnetic coils. After Jedlik solved the technical problems of continuous rotation with the invention of the commutator, he called his early devices "electromagnetic self-rotors". Although they were used only for teaching, in 1828 Jedlik demonstrated the first device to contain the three main components of practical DC motors: the stator, rotor and commutator. The device employed no permanent magnets, as the magnetic fields of both the stationary and revolving components were produced solely by the currents flowing through their windings.

An electric motor is an electrical machine that converts electrical energy into mechanical energy. Most electric motors operate through the interaction between the motor's magnetic field and electric current in a wire winding to generate force in the form of rotation of a shaft. Electric motors can be powered by direct current (DC) sources, such as from batteries, motor vehicles or rectifiers, or by alternating current(AC) sources, such as a power grid, inverters or electrical generators. An electric generator is mechanically identical to an electric motor, but operates in the reverse direction, converting mechanical energy into electrical energy.

Electric motors may be classified by considerations such as power source type, internal construction, application and type of motion output. In addition to AC versus DC types, motors may be brushed or brushless, may be of various phase(see single-phase, two-phase, or three-phase), and may be either air-cooled or liquid-cooled. General-purpose motors with standard dimensions and characteristics provide convenient mechanical power for industrial use. Figures 14 – 4, 14 – 5 and 14 – 6 show some typical schematic diagrams. The largest electric motors are used for ship propulsion, pipeline compression and pumped-storage applications with ratings reaching 100 megawatts. Electric motors are found in industrial fans, blowers and pumps, machine tools, household appliances, power tools and disk drives. Small motors may be found in electric watches.

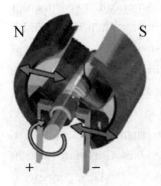

Figure 14 – 4　A brushed DC electric motor

In certain applications, such as in regenerative

braking with traction motors，electric motors can be used in reverse as generators to recover energy that might otherwise be lost as heat and friction[2].

Electric motors produce linear or rotary force(torque) and can be distinguished from devices such as magnetic solenoids and loudspeakers that convert electricity into motion but do not generate usable mechanical force，which are respectively referred to as actuators and transducers.

Figure 14－5　Cutaway view through stator of induction motor

Figure 14－6　Electric motor rotor (left) and stator(right)

Words and Expressions

electrostatic	[ˌɪˌlektrə(ʊ)ˈstætɪk]	adj.	静电的;静电学的
theoretical	[θɪəˈretɪkl]	adj.	理论的;理论上的;假设的;推理的
persistent	[pəˈsɪstənt]	adj.	执着的,坚持不懈的;持续的,反复出现的
interaction	[ɪntərˈækʃn]	n.	相互作用,相互影响;交流
electromagnetic	[ɪˌlektrə(ʊ)mægˈnetɪk]	adj.	电磁的
demonstration	[demənˈstreɪʃn]	n.	示范;证明;示威游行
dip	[dɪp]	v.	浸,蘸;(使)下降,下沉,减少
brine	[braɪn]	n.	卤水;盐水;海水
mercury	[ˈmɜːkjəri]	n.	[化]汞,水银,[天]水星
refinement	[rɪˈfaɪnmənt]	n.	改进,改善;[化工]提纯
homopolar	[ˌhəʊmə(ʊ)ˈpəʊlə]	adj.	同极的;[电]单极的
commutator	[ˈkɒmjuˌteɪtə]	n.	[电]换向器;整流器
stator	[ˈsteɪtə]	n.	固定片,定子
rotor	[ˈrəʊtə]	n.	[电][机][动力]转子;水平旋翼;旋转体
inverter	[ɪnˈvɜːtə]	n.	换流器;[电子]反相器

| brushless | [ˈbrʌʃləs] | *adj.* 无刷,无刷的 |
| blower | [ˈbləʊə] | *n.* 鼓风机,吹风机;吹制工 |

NOTES

[1] It only took a few weeks for Andre Marie Ampere to develop the first formulation of the electromagnetic interaction and present the Ampere's force law, that described the production of mechanical force by the interaction of an electric current and a magnetic field.

安德烈·玛丽·安培只花了几周的时间就发现了电磁相互作用的第一个公式,并提出了电磁力定律,描述了电流和磁场相互作用产生的机械力。

[2] In certain applications, such as in regenerative braking with traction motors, electric motors can be used in reverse as generators to recover energy that might otherwise be lost as heat and friction.

在某些应用中,如在牵引电机的再生制动中,电动机可以作为发电机反向使用,以回收通过热量和摩擦形式损失的能量。

阅读材料

电动机

第一台电动机是简单的静电装置,由苏格兰僧侣安德鲁·戈登和美国实验学家本杰明·富兰克林在18世纪40年代的实验中完成测试。其中的理论原理(库仑定律)由亨利·卡文迪什在1771年发现。由于在当时没有公开出版,1785年,查尔斯·奥古斯丁·库仑独立发现了这条定律后,以他的名字将其命名为库仑定律并公开出版。1799年亚历山德罗·沃尔特发明的电化学电池使持续电流的产生成为可能。1820年,汉斯·克里斯蒂安发现了这种电流与磁场的相互作用,即电磁作用,很快此技术就取得了很大的进展。安德烈·玛丽·安培只花了几周的时间就发现了电磁相互作用的第一个公式,并提出了电磁力定律,描述了电流和磁场相互作用产生的机械力。1821年,迈克尔·法拉第首次演示了电磁旋转运动的效果,他将一根自由悬挂的金属丝浸入汞池中,在汞池上放置一块永久磁铁,当电流通过导线时,导线绕着磁铁旋转,表明电流在导线周围产生了一个闭合的圆形磁场。这种小马达经常在物理实验中用来演示,只是在演示中用盐水代替了(有毒的)水银。

1827年,匈牙利物理学家杰德利克开始用电磁线圈做实验,在杰德利克发明了换向器解决了连续旋转的技术问题后,他称他的早期装置为"电磁自转"。在1828年,尽管它们只用于教学,杰德利克演示了第一个包含实用直流电机三个主要部件的装置:定子、转子和换向器。该装置不使用永磁体,其中静止部件和旋转部件的磁场都是由流过其绕组的电流产生的。

　　电动机是将电能转化为机械能的电机。大多数电动机都是通过电动机磁场和线圈中电流的相互作用来产生使轴转动的力。电动机可以由直流电源供电,如蓄电池、汽车或整流器,或由交流电源供电,如电网、变频器或发电机。发电机在机械原理上与电动机相同,但工作方向相反,它是将机械能转化为电能。

　　电动机可根据电源类型、内部结构、应用场合和运动输出类型等因素进行分类。除交流和直流外,电机还可以分为有刷或无刷,可以是各种相位(见单相、两相或三相),可以是空冷或液冷。图 14-4、14-5、14-6 给出了一些典型的示意图。通用电动机具有标准的尺寸和特性,能为工业生产提供便捷的机械动力。最大的电动机用于船舶推进、管道压缩和抽水蓄能应用,额定功率可达 100 兆瓦。电动机还应用于工业风扇、鼓风机、泵、机床、家用电器、电动工具和磁盘驱动器中。小型电动机还应用在电子表中。

图 14-4　有刷直流电机

图 14-5　感应电机定子的剖视图

图 14-6　电机转子(左)和定子(右)

　　在某些应用中,如在牵引电机的再生制动中,电动机可以作为发电机反向使用,以回收通过热量和摩擦形式损失的能量。

　　电动机产生线性或旋转力(扭矩),这和电磁阀及扬声器的原理有所区别,后者是将电能转换为动能但不产生可用的机械力,进而分别被称为执行器和变频器。

Unit 15

Plastics Forming and Molds

1. General

The continuing development of injection mold technology demands more and more of the processes. The most important problem in the process of injection molding is undoubtedly the correct design of injection mold, because the molding shop has little influence, if any, on the construction of the machine[1]. Efficient production of the most diverse injection-molded parts depends primarily on the injection mold.

The durability of the molds depends on their care and treatment. Since the moving components and cavity of the mold are always hardened and ground, they should produce between 500 000 and 1 000 000 000 shots.

For ease of construction and to lower manufacturing cost, injection molds are becoming standardized. Some firms offer ready-made bases of square or round design as standard or stripper plate molds for immediate use. Only the inserts then have to be fitted into the bases.

2. Basic Mold Construction

(1) Operating Principle

Basically the injection mold consists of two halves. One mold half contains the sprue bushing and runner system, the other half houses the ejection system. The molded part is located at the parting line(ref. To DIN 16 700).

(2) Single-or Multi-cavity Molds

To set up a calculation conceiving the choice of cavities in an injection mold requires accurate knowledge of the material to be processed, of the injection-molding machine and of the molds. The mold costs increase with the rising number of cavities and the relative machine costs decrease. The production time required for a given molded part depends on the wall thickness, the injection speed, the recovery rate, the time required to cool the molded material, the cooling capacity of the mold and the necessary incidental time such as duration of pressure holding time, ejection time, delay time, etc.

Therefore the decision concerning the number of cavities to be determined depends on:

① size of the order(number of molded parts in connection with delivery time);

② shape of the molded parts(size, quality requirements);

③ injection-molding machine (clamping force, plasticizing and injection capacity);

④ mold costs.

There are several known procedures for calculating the economical number of cavities. Unfortunately they are so varied that it is impossible to condense them.

As an example:

The theoretical max number of cavities is:

F1 = Max injection volume of machine(cm^3)/Volume of molded parts, sprue and runner(cm^3)

F2 = Plasticizing capacity(cm^3/min)/Number of shots/min × [part, sprue and runner(cm^3)]

F2 must be equal to or smaller than F1.

F1 is the maximum number of cavities attainable.

3. Three-plate Injection Molds

Three-plate injection molds are injection molds possessing two parting lines, having been designed as degating molds. This means that the sprue is separated from the parts in the mold. At the conclusion of injection the mold opens at the first parting line, which contains the sprue and runner and separated the pinpoint gate from the part. Only when the second parting line opened, the part would be released. There are several possibilities for achieving this.

In Figure 15 - 1, with short stroke distances of plate 2, compression spring 4 or spring washers can be inserted between plate 1 and plate 2, which push plate 2 forward. The stop 3 limits the stroke.

The retention system of plate 1 shown in Figure 15 - 2 is often employed. Dowel 2 is provided with an annular groove 3 into which the ball catch 4 is pushed. A not too heavy plate can be carried along by the resultant adhesive effect.

The plate 2 to be moved in Figure 15 - 3 is connected to the ejector plates by ejector rods, which in turn are fixed in their basic position by strong spring 3. Plate 2 is carried along until the ejector plates are stopped by running against the ejector bar of the machine. Then only the complete mold body keeps moving on in the usual manner. Care must be taken with this design that

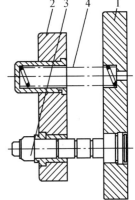

Figure 15 - 1   Spring type
1—plate;2—push plate;3—stop;
4—compression spring

the ejector bar 5 is slightly delayed.

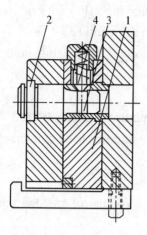

Figure 15 - 2　Retention system
1—plate；2—dowel；3—annular groove；
4—ball catch

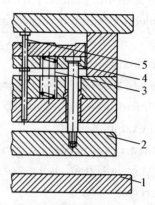

Figure 15 - 3　Three-plate injection mold
1—plate 1；2—plate 2；3—spring；
4—ejector rod；5—ejector bar

They must not be rigidly connected to the ejector plate；otherwise there would be a tendency for the parts to be pushed back into the cavities in plate 2. With this design the stroke of plate 2 can be arbitrary. Moreover，when the ejector rods 4 are hardened and guided in bushes，these ejector rods can take over the function of the guide pins fitted on the fixed mold half，so that the space between plates 1 and 2 is not cluttered up with protruding rods or springs. This can be of advantage if a rather complicated runner system has to be designed.

Figure 15 - 4 shows a mechanical type of plate movement. Although more elaborate，it is nevertheless the safest way，particularly if large，heavy plates have to be moved. The latch 3，which hooks over plate 1，moves the latter along. The movement is continued until the pin 2 in the hook is lifted by the rise in the guide strip 4 on which it runs. The pin has to escape upward，thereby releasing the hook from the plate 2. The plate now pushes up against a stop，which in this case has been fitted to the underside of the guide strip at 6，and the mold can continue opening. The hook is pushed down again by the leaf spring 5.

Figure 15 - 5 illustrates a typical degating tool. Contrary to the usual mold design，this tool has an additional mold parting line for the star-shaped runner[2]. Mold constructed on this principle is preferably employed because of the simplicity of the gating technique. But there is some concern that bulky runner often do not drop easily between the mold parting lines. It is important that a sufficiently large gap always be set.

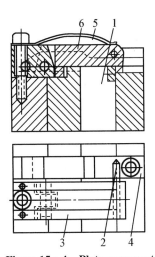

Figure 15-4　Plate movement

1—plate;2—pin;3—latch;4—guide strip;
5—leaf spring;6—waterway

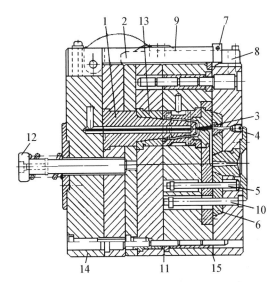

Figure 15-5　Typical degating tool

1—sprue separating area;2—latch;3—sprue;4—retainer pin;
5—head of bolt;6—sprue-stripping plate;7—pin;
8—guide strip; 9—waterway;10—bolt;
11—main parting line;12—ejector rod; 13—stripper sleeve;
14—plate assembly;15—mold cores

At the conclusion of injection the mold first parted at 1 in the sprue separating area when the machine opens[3]. The complete mold block, with the exception of the fixed mold half, is interlocked within itself by the latch 2. As the sprue 3 is held captive by the undercut in the retainer pin 4, the part is severed from the sprue, which stays on the fixed half.

When the head of the bolt 5 catches the sprue-stripping plate 6, this is moved forward,thereby stripping the sprue from the retaining pin 4. When the sequence has been completed, the pin 7 of latch 2 has arrived at the inclined plane of the guide strip 8 in position 9, forcibly lifting the latch and thereby ending the interlock.

The head of bolt 10 reached the plates along the fixed position. The continuing sequence of the opening movement consists of tile mold opening at the main parting line 11, thereby creating the drop-out opening for the ejected part. As the opening sequence continues, the head of ejector rod 12 pushes against the ejector of the molding machine and stops the plate assembly 14 as well as the stripper sleeve 13. The mold cores 15 fitted into the plate assembly 14 withdraws from the part, and the ejection process is concluded.

Words and Expressions

injection mold			注射模具
molding shop			成型车间,造型车间
injection-molded			注射成型的
grind(ground,ground)			磨削,磨光
single-cavity mold			单型腔注射模具
multi-cavity mold			多型腔注射模具
clamping force			夹紧力,合模力
plasticize	['plæstɪsaɪz]	vt.	(使)成为可塑;(使)可塑
injection capacity			注射容量
three-plate injection mold			三板式注射模具
sprue	[spruː]	n.	注料口,流道,直浇道
pinpoint gate			针孔型浇口
compression spring			压缩弹簧
spring washer			弹簧垫圈
stop	[stɒp]	n.	止动销,制动销,限位器
retention system			滞留系统
dowel	['daʊəl]	n.	[机]定位销,轴销
annular groove			环形凹槽
ball catch			门碰球,球掣
ejector plate			顶料板
ejector rod			顶料杆,出料杆
protruding rod			凸出杆,伸出杆
protruding spring			凸出弹簧
latch	[lætʃ]	n.	插销,撞锁,弹簧锁
pin	[pɪn]	n.	栓,钉,销
leaf spring			片弹簧
star-shaped runner			星形流道
gating technique			脉冲选通技术
bulky runner			大流道,大浇道
sprue separating area			浇道分离面
undercut	[ˌʌndəˈkʌt]	n.	下(部凹)陷
retainer pin			开口销,固定销
sprue-stripping plate			注料口分模板
stripping			抽钉,脱模,拆模

inclined plane　　　　　　　　　　倾斜面,斜面

plate assembly　　　　　　　　　　板装置,板组合

stripper sleeve　　　　　　　　　　分型(模)套

mold core　　　　　　　　　　　　型芯

NOTES

［1］The most important problem in the process of injection molding is undoubtedly the correct design of injection mold，because the molding shop has little influence，if any，on the construction of the machine.

注射成型过程中最重要的问题毫无疑问是注射模具的正确设计,因为成型车间对注射件的影响很小,即便有,也是对机器的结构有影响的。(句中"if any"引出转折状语从句,译为"即便有")

［2］Contrary to the usual mold design，this tool has an additional mold parting line for the star-shaped runner.

与通常的模具设计不同,在此模具中为星形流道设计了一个附加的模具分型线。(句中"Contrary to"译为"与……相反")

［3］At the conclusion of injection the mold first parted at 1 in the sprue separating area when the machine opens.

当注射成型结束时,随着机器的打开,模具将首先在浇道分离面1处分型。(句中短语"At the conclusion of"译为"当……完结时")

第 15 单元　塑料成型与模具

1. 概要

注射模技术的不断发展需要越来越多的工艺流程。注射成型过程中最重要的问题毫无疑问是注射模具的正确设计,因为成型车间对注射模具的影响很小,即便有,也是对机器的结构有影响的。高效率地生产各类注射成型零件主要取决于注射模具。

模具的使用寿命取决于对它们的维护保养和加工处理,例如对一些活动件及型腔的表面硬化和磨削加工,使得它们的生产批量可以达到 500 000～1 000 000 000 件。

为了简化模具结构和降低制造成本,使注射模具标准化,许多制造商提供了已经生产好的标准方形或圆形底及脱模板供直接使用,仅需要将它们很好地与机座配合起来。

2. 模具的基本结构

（1）工作原理

从根本上说,注射模具包括两个部分:一部分包括浇口和浇注系统,另一部分是放置顶出系统。模具零件是在分型面被定位的(参考 DIN 16700)。

（2）单型腔注射模具及多型腔注射模具

注射模具型腔选择等设计要求掌握加工材料、注射机和模具等方面的准确知识。模具的制造成本随着型腔数目的增加而增加，而相关的加工费用减少了。一个给定的模具零件的生产周期取决于壁厚、注射速度、收缩率、模内材料的冷却时间、冷却的效能及必要的辅助时间，如压力持续时间、排气时间及延迟时间等。

因此确定型腔数量的主要因素涉及以下几点：

① 订单大小（制品数量及相关交货时间）；

② 制品形状（尺寸、质量要求）；

③ 注射机（合模力、可塑性、注射容量）；

④ 模具费用。

尽管已有多种计算经济型腔数目的方法，但这些计算方法多种多样，根本没法将它们提炼成为简便公式。

例如：

型腔的理论最大数目为

F1＝注射机注射容量的最大值（cm³）/制件、注料口和流道体积（cm³）

F2＝注塑容量（cm³/min）/制造数量/min[制件、注料口和流道（cm³）]

F2必须小于等于F1，F1是可得到的型腔数目的最大值。

3. 三板式注射模具

三板式注射模具是具有两个分型面的注射模具，这种模具带有浇口，也就是直浇道在模型中和塑件相分离。在注射结束之后，模具沿第一层分型面开模，其中包含了注料口、浇道和塑件中分流的针孔型浇口。仅当沿第二层分型面开模时，塑件才脱离型芯。有多种可能性来实现这种情况。

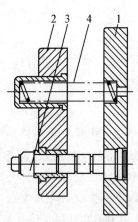

图15－1 弹簧式

1—模板；2—动模板；
3—止动销；4—压缩弹簧

在图15－1中，模板2具有一个短小的行程距离，压缩弹簧4或弹簧垫圈可以插入模板1和模板2之间，推动模板2向前，而止动销3则可以限制模板2的行程。

在图15－2中，模板1在滞留系统中也经常被使用。定位销2中有一个环形凹槽3可以使球掣4进入，一块不太厚的模板可以沿着黏合面而移动。

图15－3中的模板2通过顶料杆和顶料板相连。顶料杆由强度很大的弹簧3依次固定在它们的基本位置。直到顶料板撞上机器的推出杆，模板2的移动才停止。当一个完整的模型保持正常方式运行时，需注意推出杆5应延时推出。

顶料杆不能和顶料板刚性相连，否则塑件将会有被推回模板2型腔的可能。对于这种设计，模板2的行程可以是任意的。而且当顶料杆4被加固并且用轴套引导时，这些顶料杆可替代导向销的作用来与定模部分配合，所以在模板1和模板2之间不能乱放伸出杆或弹簧。如果要设计一个非常复杂的浇道系统，这种方法具有优势。

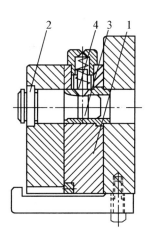

图 15-2 滞留系统

1—模板;2—定位销;
3—环形凹槽;4—球掣

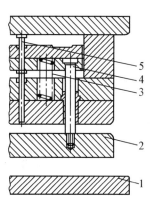

图 15-3 三板式注射模具

1—模板1;2—模板2;3—弹簧;
4—顶料杆;5—推出杆

图 15-4 所示为平板运动的机械形式。虽然这种形式更为精细,但这无疑是最安全的方法,特别是当大而重的平板需要被移动时。挂在模板 1 上的插销 3 沿着模板移动,直到销被滑块抬起,销不得不向上移动,所以释放了模板 2 上的挂钩。这时模板反向止动销向上推,在这种情况下,止动销已在 6 处与导轨的下侧相配合,所以模具能一直打开。然后,挂钩再一次被片弹簧 5 向下拉。

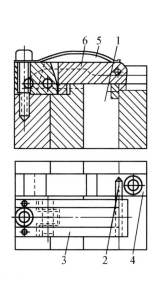

图 15-4 模板运动

1—模板;2—销;3—插销;3—直浇道;
4—导轨;5—片弹簧;6—冷却水道

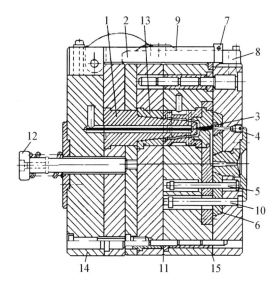

图 15-5 典型注塑模具

1—浇道分离面;2、10—插销;3—浇道;4—固定销;
5、10—插销头;6—注料口分模板;
7—销;8—导轨;9—冷却水道;
11—主要分型线;12—预料杆;
13—分型(模)套;14—模板装置;
15—模具型芯

图 15-5 所示为一个典型注塑模具。与通常的模具设计不同,在此模具中为星形流道设计了一个附加的模具分型线。因为有简单的脉冲选通技术,所以用这种方法设计的模具更适于使用。但是需要注意的是,通常大浇道不易滴在分型线之间,所以大的间隙设计是非常重要的。

当注射成型结束时,随着机器的打开,模具将首先在浇道分离面 1 处分型。一个完整的具有非固定板的模型装置是通过插销 2 使它们连接在一起,直浇道 3 被固定销 4 中的下陷所约束。塑件从直浇道被切断,直浇道继续停留在固定部分。

当插销头 5 的销头到达注料口分模板 6 时,这次运动停止,从而直浇道从止动销 4 中脱模。当运作完成时,插销 2 中的销 7 在位置 9 处沿着导轨 8 到达倾斜面,强制抬起插销从而结束互锁。

插销 10 的销头沿着固定方向到达模板。开模运动的持续进行由塑料模型在主要分型线 11 处的开模组成,从而为喷出部分创造了脱落开模。当开模动作继续进行时,顶料杆 12 的杆头推动模具机构的推出器,使板装置 14 和分型套 13 停止。装在板装置 14 中的型芯 15 从塑件中退出,从而注射过程结束。

READING MATERIAL

Pressure Die Casting

Pressure die casting is the process of forcing molten metal under high pressure into mold cavities(which are machined into dies). Most pressure die castings are made from nonferrous metals, specifically zinc, copper, aluminum, magnesium, lead, and tin based alloys, but ferrous metal pressure die castings are possible. The pressure die casting method is especially suited for applications where a large quantity of small to medium sized parts is needed with good detail, a fine surface quality and dimensional consistency. This process, similar to injection molding, is a variant of porous mold casting in which the ceramic suspension is injected into the mold under high pressure[1]. The molds may be fabricated from plastic, plaster or ceramics. And the higher the applied pressure, the shorter the casting time.

There are four major steps in the pressure die casting process. Firstly, the mold is sprayed with lubricant and closed. The lubricant both helps control the temperature of the die and it also assists in the removal of the casting. Secondly, the molten metal is then shot into the die under high pressure; between 10~175 MPa (1 500~25 000 psi). Once the die is filled, the pressure is maintained until the casting has solidified. Then, the die is opened and the shot(shots are different from castings because there can be multiple cavities in a die, yielding multiple castings per shot) is ejected by the ejector pins. Finally, the scrap, which includes the gate, runners, sprues and flash, must be separated from the casting(s). This is often done using a special trim die in a power press or hydraulic press. An older method is separating by hand or by sawing, which case grinding may be necessary to smooth the scrap marks. A less labor-intensive method is to tumble shots if gates are thin and easily broken; separation of gates from finished parts must follow. This scrap is recycled by remelting it.

The high-pressure injection leads to a quick fill of the die, which is required so the entire cavity fills before any part of the casting solidifies. In this way, discontinuities are avoided even if the shape requires difficult-to-fill thin sections. This creates the problem of air entrapment, because when the mold is filled quickly there is little time for the air to escape. This problem is minimized by including vents along the parting lines, however, even in a highly refined process there will still be some porosity in the center of the casting.

Most die casters perform other secondary operations to produce features not readily castable, such as tapping a hole, polishing, plating, buffing, or painting.

Pressure die casting has some advantages and disadvantages as follows.

Advantages:

(1) Excellent dimensional accuracy(dependent on casting material).

(2) Smooth cast surfaces.

(3) Thinner walls can be cast as compared to sand and permanent mold casting approximately 0.75 mm.

(4) Inserts can be cast-in(such as threaded inserts, heating elements, and high strength bearing surfaces).

(5) Reduces or eliminates secondary machining operations.

(6) Rapid production rates.

(7) Casting tensile strength as high as 415 MPa.

Disadvantages:

(1) Casting weight must be between 30 grams and 10 kg.

(2) Casting must be smaller than 600 mm.

(3) High initial cost.

(4) Limited to high-fluidity metals.

(5) A certain amount of porosity is common.

(6) Thickest section should be less than 13 mm.

(7) A large production volume is needed to make this an economical alternative to other processes.

The main pressure die casting alloys are: zinc, aluminum, magnesium, copper, lead, and tin. The following is a summary of the advantages of each alloy:

(1) Zinc: the easiest alloy to cast; high ductility; high impact strength; easily plated; economical for small parts; promotes long die life.

(2) Aluminum: light weight; high dimensional stability for complex shapes and thin walls; good corrosion resistance; good mechanical properties; high thermal and electrical conductivity; retains strength at high temperatures.

(3) Magnesium: the easiest alloy to machine; excellent strength-to-weight ratio; lightest alloy commonly die cast.

(4) Copper: high hardness; high corrosion resistance; highest mechanical properties of alloys die cast; excellent wear resistance; excellent dimensional stability; strength approaching that of steel parts.

(5) Lead and tin: high density; extremely close dimensional accuracy; used for special forms of corrosion resistance.

Maximum weight limits for aluminum, brass, magnesium, and zinc castings are approximately 32 kg, 5 kg, 20 kg, and 34 kg, respectively.

The pressure die casting die is an assembly of materials, mostly ferrous metals, each of which plays a part in a mechanism which will operate under conditions of rapidly changing temperatures, as the molten metal is injected under pressure and then immediately cooled[2].

Words and Expressions

pressure die casting			压力铸造
dimensional	[daɪˈmenʃnəl]	adj.	空间的,尺寸的
injection molding			注射成型
variant	[ˈveəriənt]	n.	变量
porous	[ˈpɔːrəs]	adj.	多孔渗水的
ceramic	[səˈræmɪk]	n.	陶瓷制品
fabricate	[ˈfæbrɪkeɪt]	vt.	制作,构成
plaster	[ˈplɑːstə]	n.	石膏
spray	[spreɪ]	vt.	喷射,喷溅
lubricant	[ˈluːbrɪkənt]	n.	润滑剂
solidify	[səˈlɪdɪfaɪ]	v.	(使)凝固,(使)团结,巩固
ejector pin			推杆
scrap	[skræp]	n.	废料
gate	[ɡeɪt]	n.	浇口
runner	[ˈrʌnə]	n.	分流道
sprue	[spruː]	n.	主流道
flash	[flæʃ]	n.	飞边
trim die			剪修模

hydraulic press			液压压力机
difficult-to-fill			难以充填
polish	['pɒlɪʃ]	vi.	发亮,变光滑
plating	['pleɪtɪŋ]	n.	电镀
buffing	['bʌfɪŋ]	n.	抛光
permanent mold casting			金属铸型
insert	[ɪn'sɜːt]	n.	型芯

NOTES

〔1〕This process，similar to injection molding，is a variant of porous mold casting in which the ceramic suspension is injected into the mold under high pressure.

这个工艺和注塑成型工艺比较相似,和多孔型铸造的不同在于陶瓷熔液在高压下注入模腔中。(句中 similar to injection molding 是插入语,是"和注塑成型工艺相似"的意思)

〔2〕The pressure die casting die is an assembly of materials，mostly ferrous metals，each of which plays a part in a mechanism which will operate under conditions of rapidly changing temperatures，as the molten metal is injected under pressure and then immediately cooled.

压力铸造模具多数是由一些铁质金属材料的部件组装而成的,且每一个部件都在如下机制中承担一定作用。当在一定的压力下熔化了的液态金属被注入模具型腔且随即被冷却时,这些部件便可以在温度快速变化的条件下工作。(本句中 mostly ferrous metals 是指 materials,第一个 which 也是指 materials,第二个 which 引导的定语从句修饰 mechanism)

阅读材料

压力铸造

压力铸造是迫使熔融金属在高压下进入模腔(加工成模具)的一种工艺。大多数的压力铸件都由非铁金属制成,尤其是锌、铜、铝、镁、铅以及锡合金,但是铁质金属进行压铸也是可能的。压力铸造工艺方法特别适用于大批量生产的,需要有细节优良、表面质量和空间连贯性良好的中小尺寸的零件。这个工艺和注塑成型工艺比较相似,和多孔型铸造的不同在于陶瓷熔液在高压下注入模腔中。这个模腔可以由塑料、石膏或陶瓷制成。在铸造过程中,使用的压力越高,压铸的时间越短。

压力铸造工艺过程有四个主要的步骤。首先,模腔要用润滑剂喷射然后再闭合。

润滑剂不仅可以帮助控制模具温度,并且可以辅助移动铸件。其次,熔融的金属在高压下射入模具,压力在 10 兆帕到 175 兆帕之间。一旦模腔被填满,一定要保持压力直到铸件固化。再次,模具打开,注射点被推杆推出(注射点不同于铸件,因为一个模具中可以有多个模腔,每个注射点可产生多个铸件)。最后,浇口、分流道、主流道和飞边等废料必须和铸件分离。这个步骤通常使用专门的动力压力机或液压压力机的剪切模具来完成。当必须要进行表面研磨来磨平废边印痕的时候,一个古老的方法是采用手工或锯的方法把废料和铸件分离。一个劳动强度较小的方法是如果浇口很细且容易破裂则可以摔掉注射点,而紧接着必须把浇口从成品上分离。这个废料可以通过再熔融回收。

高压注射导致了充模迅速,这使得在铸件的任何一个部分固化之前模腔完全被充满。这样,即使是很难充填的细小区域也可以充填,避免了铸件的不连续。这会引起空气残留的问题,因为当模具被快速充满的时候,只有很少的时间让空气排出。这个问题可以通过在分型面上扎通气孔而最小化,然而,即使是在非常精密的工艺过程中,在铸件的中心还是会存在一些气孔。

大多数铸造师还要做第二道工序以制造并不容易铸造的铸件,比如说打孔、研磨、电镀、抛光或着色。

压力铸造具有如下一些优缺点。

优点:

(1) 非常好的尺寸精度(根据铸造材料)。

(2) 光滑的铸造表面。

(3) 和砂型铸造及金属铸型铸造相比可以铸造大约 0.75 毫米的薄壁铸件。

(4) 可以铸造型芯(比如细长型芯、发热元件和高强度支承面)。

(5) 减少或消除二次加工操作。

(6) 高的生产率。

(7) 铸件抗拉强度可以高达 415 兆帕。

缺点:

(1) 铸件的重量必须在 30 克到 10 千克之间。

(2) 铸件尺寸必须小于 600 毫米。

(3) 最初成本较高。

(4) 限于铸造高流动性金属。

(5) 通常伴有一定量的气孔。

(6) 铸件厚度需小于 13 毫米。

(7) 大批量的生产才使得铸造工艺比较经济地替代其他工艺。

主要的压力铸造合金包括:锌、铝、镁、铜、铅和锡。以下是每种合金的优点摘要:

(1) 锌:最容易铸造的合金,高延展性、高抗冲击强度、易电镀、铸造小零件比较经济,可延长模具寿命。

(2) 铝:重量轻,对于形状复杂和壁薄的铸件能保持良好的稳定性,耐腐蚀,高机械性能,高热电导率,在高温下能保持强度。

(3) 镁:最容易切削的合金,极好的比强度,压铸件中最轻的合金。

(4) 铜:高硬度,耐腐蚀,具有合金压铸件中最好的机械性能,极好的耐磨损性,极好的尺寸稳定性,强度接近钢制品。

(5) 铅和锡:高密度、非常精密的尺寸精度,可用于铸造耐腐蚀的专门制品。

对于铝铸件、铜铸件、镁铸件和锌铸件的最大重量限制分别是 32 千克、5 千克、20千克和 34 千克。

压力铸造模具多数是由一些铁质金属材料部件组装而成,且每一个部件都在如下机制中承担一定作用。当在一定的压力下熔化了的液态金属被注入模具型腔且随即被冷却时,这些部件便可以在温度快速变化的条件下工作。

Unit 16

Stamping Forming and Die Design

扫码可见本项目
参考资料

1. Stamping Forming

Metal processing is a branch of engineering science，which deals with the manufacturing of metallic parts and structures through the processes of plastic forming，machining，welding and casting. This part focuses on the stamping forming technology and its die design in metal processing. Stamping is mainly used in sheet plate forming，which can be used not only in metal forming，but also in non-metal forming. In stamping forming，under the action of dies，the inner force deforming the plate occurs in the plate. When the inner force reaches a certain degree，the corresponding plastic deformation occurs in the blank or in some region of the blank. Therefore the part with certain shape，size and characteristic is produced.

Stamping is carried out by dies and press，and has a high productivity. Mechanization and automatization for stamping can be realized conveniently owing to its easy operation. Because the stamping part is produced by dies，it can be used to produce the complex part that may be manufactured with difficulty by other processes. The stamping part can be used generally without further machining. Usually，stamping process can be done without heating. Therefore，not only does it save material but also energy. Moreover，the stamping part has the characteristics of light weight and high rigidity.

Stamping processes vary with the shape，the size and the accuracy demands and the material. It can be classified into two categories：cutting process and forming process. The objective of cutting process is to separate the part from blank along a given contour line in stamping. The surface quality of the cross-section of the separated part must meet a certain demand. In forming processes，such as bending，deep drawing，local forming，bulging，flanging，necking，sizing and spinning，plastic deformation occurs in the blank without fracture and wrinkle，and the part with the required shape and dimensional accuracy is produced.

The stamping processes widely used are listed in Table 16 – 1.

Table 16 - 1   Classification of the stamping processes and their characteristics

Process		Diagram	Characteristics
Cutting	Punching		Separate the blank along a closed outline; the cut down part is waste
	Blanking		Separate the blank along a closed outline; the cut down part is workpiece
	Lancing		Partly separate the blank along a unclosed outline, bending occurs at the separated part
	Shearing		Shear the plate into strip or piece
	Parting		Separate various workpiece produced by stamping into two or more parts
Forming	Bending		Press the sheet metal into various angles, curvatures and shapes
	Curling		Bend ending portion of the plate into nearly closed circle
	Deep drawing		Produce an opened hollow part with punch and die
	Local forming		Manufacture various convex or concave on the surface of the plate or part

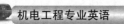
Process		Diagram	Characteristics
Forming	Bulging		Expand a hollow or tubular blank into a curved surface part
	Flanging		Press the edge of the hole or the external edge of the workpiece into vertical straight wall
	Necking		Decrease the end or middle diameter of the hollow or tubular shaped part
	Sizing		Finish the deformed workpiece into the accurate shape and size
	Spinning		Form an axis-symmetrical hollow part by means of roller feeding and spindle rotational movement

2. Blanking and Punching Dies

（1）Simple Die

The die that only one process is carried out in one press stroke is called simple die. Its structure is simple（see Figure 16 - 1）, so it can be easily manufactured. It is applicable to small batch production.

（2）Progressive Die

The die that several blanking processes are carried out at different positions of the die in one press stroke is called progressive die, as shown in Figure 16 - 2. In the operation, the locating pin 2 aims at the locating holes punched previously, and the punch moves downwards to punch by punch 4 and to blank by punch 1, thus the workpiece 8 is produced. When the punch returns, the stripper 6 scrapes the blank 7 from the punch 4, the blank 7 moves forward one step and then the second blanking begins. Above steps are repeated continually. The step distance of the blank is controlled by a stop pin.

(3) Compound Die

The die that several processes are carried out at the same die position in one press stroke is called compound die, as shown in Figure 16 – 3. The main characteristic of the compound die is that the part 1 is both the punch and the die. The outside circle of the punch-die 1 is the cutting edge of the blanking punch, while the inside hole is a deep drawing die. When the slide moves downwards along with the punch-die 1, the blanking process is done first by the punch-die 1 and the blanking die 4, the blanked workpiece is pushed by deep

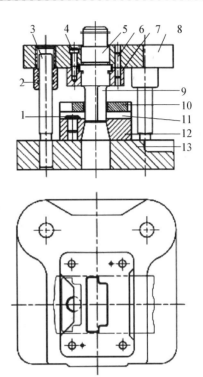

Figure 16 – 1   Simple die

1—stop pin;2—guide bushing;3—guide pin;4—bolt;
5—die shank;6—pin;7—fixed plate;8—upper bolster;
9—punch;10—stripper;11—stock guide;
12 die;13—lower bolster

drawing punch 2, and then the deep drawing die moves downwards to carry out deep drawing operation. The ejector 5 and the stripper 3 push the deep drawn workpiece 9 out of the die when the slide returns. The compound die is suitable for mass production and high accuracy blanking.

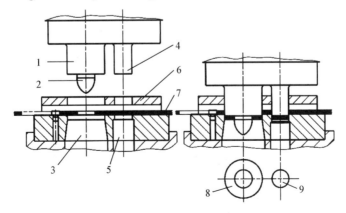

Figure 16 – 2   Progressive die for blanking and punching

1—blanking punch;2—locating pin;3—blanking die;4—punching punch;
5—punching die;6—stripper;7—blank;8—workpiece;9—waste

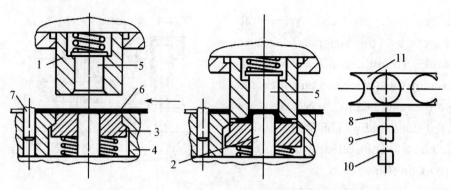

Figure 16 - 3 Compound die for blanking and deep drawing

1—punch die;2—deep drawing punch;3—press plate(stripper);4—blanking die;5—ejector;
6—strip blank;7—stop pin;8—blank;9—deep drawn workpiece;10—finished part;11—waste

Words and Expressions

stamping	['stæmpɪŋ]	n.	冲压,冲压件
die	[daɪ]	n.	模具,砧子,凹模
metal processing			金属加工
metallic engineering science			金属工程科学
plastic forming			塑性成形
machining	[mə'ʃiːnɪŋ]	n.	机械加工,切削加工
welding	['weldɪŋ]	n.	焊接
casting	['kɑːstɪŋ]	n.	铸造,浇注
plate	[pleɪt]	n.	板,板材,钢板
blank	[blæŋk]	n.	毛坯,坯料
press	[pres]	v.	冲压,压制
mechanization	[ˌmekənaɪ'zeɪʃn]	n.	机械化
automatization	[ɔːtəumətɪ'zeʃn]	n.	自动化
rigidity	[rɪ'dʒɪdəti]	n.	刚性,刚度
accuracy	['ækjərəsi]	n.	精度
cutting process			分离工序
forming process			成形过程,成形工艺
contour	['kɒntʊə]	n.	轮廓,外形
cross-section		n.	横截面
bending	['bendɪŋ]	n.	弯曲
deep drawing			深拉深
local forming			局部成形

bulging	[ˈbʌldʒɪŋ]	n.	胀形,起凸
flanging	[ˈflændʒɪŋ]	n.	翻口,翻边,弯边
necking	[ˈnekɪŋ]	n.	缩颈
sizing	[ˈsaɪzɪŋ]	n.	整形,矫正
spinning	[ˈspɪnɪŋ]	n.	旋压,赶形
blanking	[ˈblæŋkɪŋ]	n.	落料,冲裁
fracture	[ˈfræktʃə]	n.	断裂,断裂面
wrinkle	[ˈrɪŋkl]	n.	起皱
lancing	[ˈlɑːnsɪŋ]	n.	切缝,切口
shearing	[ˈʃɪərɪŋ]	n.	剪切
strip	[strɪp]	n.	条料,带料,脱模
parting	[ˈpɑːtɪŋ]	n.	剖切,分开,分离
curvature	[ˈkɜːvətʃə]	n.	弯曲,曲率
curling	[ˈkɜːlɪŋ]	n.	卷边,卷曲
punch	[pʌntʃ]	n.	冲头,冲孔
convex	[ˈkɒnveks]	n.	凸形
tubular blank			管状坯
curved-surface			曲面
tubular-shaped			管状的
axis-symmetrical			轴对称的
spindle	[ˈspɪndl]	n.	轴,主轴
simple die			单工位模
progressive die			连续模
locating pin			定位销
stripper	[ˈstrɪpə]	n.	脱模杆,卸料板
scrap	[skræp]	n.	废料,切屑
stop pin			定位销,挡料销
compound die			复合模
slide			滑块
mass production			大量生产

第16单元　冲压成形与模具设计

1. 冲压成形

　　金属加工是工程科学的一个分支,涉及通过塑性变形、机加工、焊接和铸造的工艺来进行金属零件和结构的制造。本文主要讲冲压成形技术及其模具设计。冲压主要用于板料成形,不仅可以用于金属成形,也可以用于非金属成形。在冲压成形过程中,在模具的作用下,内应力使板料发生变形。当内应力达到某种程度,在坯料或坯料的部分区域发生相应的塑性变形,从而生产出具有一定形状、尺寸和特征的产品。

冲压由模具和冲床进行冲压,冲压成形具有很高的生产率。机械化和自动化使冲压加工易于操作。因为冲压产品是由模具生产的,所以冲压加工可以用来加工其他加工方法难以加工的复杂零件。冲压产品通常不需要进一步加工就可以直接使用。通常,完成冲压加工不需要加热。因此,冲压加工不仅节约材料而且节约能量。另外,冲压产品具有轻质刚性好的特点。

冲压工艺过程随着产品的形状、尺寸、精度要求和材料的变化而变化。冲压工艺分为两类:分离工序和成形工序。分离工序是在冲压过程中使冲件与板料沿着一定的轮廓线分离,冲压件分离断面的质量要满足一定的要求。成形工序,如弯曲、拉深、起伏成形、胀形、翻边、缩口、校平、旋压,是使冲压毛坯在不被破坏、没有起皱的条件下发生塑性变形,成为所要求的成品形状,同时也达到尺寸精度方面的要求。

常见的冲压工序见表16-1。

表16-1 冲压工艺的分类及其特点

工序名称		简图	特点
分离工序	冲孔		沿封闭轮廓线冲切,冲下的部分是废料
	落料		沿封闭轮廓线冲切,冲下的部分是零件
	切舌		沿敞开轮廓线局部分离,分离部分发生弯曲
	剪切		将板料剪切成条料
	剖切		将冲压成的半成品切开,成为两个或多个零件

续　表

工序名称		简图	特点
成形工序	弯曲		将金属板材压成各种角度、曲率和形状
	卷圆		将板材端部卷成接近封闭的圆头
	拉深		用模具将板材加工成空心零件
	起伏		在板料或零件的表面上制成各种凸起或凹陷
	胀形		将空心或管状毛坯扩张成曲面零件
	翻边		将零件的孔边缘或外边缘翻出竖立的直边

续　表

工序名称		简图	特点
成形工序	缩口		减小空心或管状零件的端部或中间直径
	校平、整形		将变形的工件校成精确的形状和尺寸
	旋压		用旋轮和主轴旋转运动形成轴对称的空心零件

2. 落料冲孔模

（1）单工序冲裁模

在压力机的一次行程中，只完成一道工序的模具称为单工序冲裁模。单工序模结构简单（如图16-1），所以容易加工制造，适合小批量生产。

（2）级进冲裁模

级进冲裁模是在压力机一次行程中，在模具的不同位置完成两个或两个以上冲裁工序的模具，如图16-2所示。冲压过程中，导正销2先导入前次冲裁冲好的导正孔，凸模下行，用凸模4冲孔，用凸模1落料，得到工件8。当凸模回程，卸料板6把板料7从凸模4上刮下，板料7向前进一个工步，再进行第二次冲裁。以上步骤不断重复。工步的距离由挡料销控制。

（3）复合冲裁模

在压力机一次冲压行程中，在模具的同一位置完成多道冲裁工序的模具称为复合模，如图16-3所示。复合模的一个主要特点是有一个零件既是凸模又是凹模。凸凹模1的外圆为落料凸模的刃口，而内孔为拉深凹模。当压力机滑块随凸凹模1向下移动时，首先由凸凹模1和落

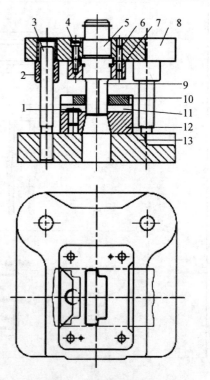

图 16-1　单工序模

1—挡料销；2—导套；3—导柱；4—螺钉；
5—模柄；6—销；7—凸模固定板；8—上模座；
9—凸模；10—卸料板；11—导料板；
12—凹模；13—下模座

料凹模4完成落料过程,然后由拉深凸模2推动工件,再由拉深凹模向下移动完成拉深
工艺过程。当滑块回程时,推件块5和卸料板3将拉深工件9推出凸凹模。复合模适
用于大批量生产和精度要求高的产品。

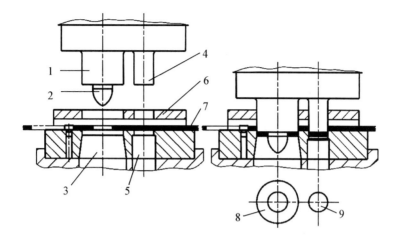

图16-2　落料冲孔级进模

1—落料凸模;2—导正销;3—落料凹模;4—冲孔凸模;5—冲孔凹模;
6—卸料板;7—坯料;8—工件;9—废料

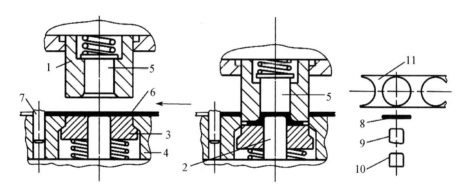

图16-3　落料拉深复合模

1—凸凹模;2—拉深凸模;3—压板(卸料板);4—落料凹模;5—推件块;6—条料;
7—挡料销;8—坯料;9—拉深件;10—成品;11—废料

READING MATERIAL

Fine Blanking

Fine blanking is a technique used for production of blanks perfectly flat and
with a cut edge which is comparable to a machined finish. This quick and easy
process is worthy of serious thought when the number of parts justifies the cost of a

blanking tool especially when consideration is given to the fact that operations such as shaving are eliminated.

Two methods of fine blanking are practiced according to the type of work undertaken as follows:

(1) The punch has a round edge and a small clearance and this is best used for blanks, but appears to give less satisfactory results when used for producing holes[1].

(2) This method uses a cutting punch which is surrounded by a V-shaped gripper which makes a V-groove round the profile and thus clamps the blank securely before a non-rounded punch contacts the surface.

In method 1 the radius on the edge of a die is selected according to the type hardness and thickness of a particular material coupled also to the shape of a profile on the component. However some importance is attached to this radius, otherwise a large increase can cause distortion and create a burr on the underside of the blank. Thus the minimum radius that will impart a good result on a component is an essential feature and this radius can vary from 0.3 to 2 mm according to conditions.

Generally the load required for fine blanking is no more than 10% greater than the load needed for conventional blanking. Greater radius and thus bursting pressures are experienced with this operation, so attention to the design of the tool is necessary and this usually means sinking the die completely into a bolster of adequate proportions.

Well-supported punches are essential to ensure that the deflection is eliminated or it will be a failure, we should pay attention to this point in a design. Sharp cutting edges are yet another item that must receive continued thought and there is apparently a case here for punches made from high-carbon high chromium steel.

The question of punch and die clearances is vital point with this design of tool and they are always much closer than those used for conventional blanking tools. As a general guide, a total clearance of 0.01 to 0.03 mm will give good results and emphasis is made of the point that these are total clearance and not each side of a hole or blank.

Stripping forces to remove the blank from the punch are higher, due to the small clearances and the load needed to strip the materials estimated as being 10% higher than normal blanking tools.

Method 2 employs the V-shaped surround to hold the blank during the operation and this projection is forced into the material. Figure 16 – 4 shows briefly what happens as the press descends; the illustrations indicate how the pressure plate clamps the strip, how blanking is carried out and how a part is ejected finally by the lower ejector.

An appreciation of the difference between ordinary blanking and fine blanking

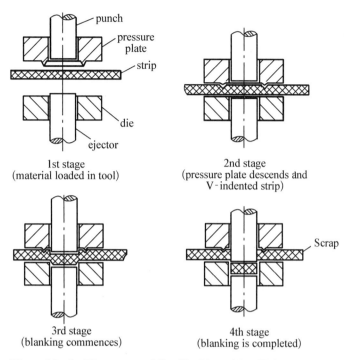

1st stage
(material loaded in tool)

2nd stage
(pressure plate descends and
V-indented strip)

3rd stage
(blanking commences)

4th stage
(blanking is completed)

Figure 16 – 4   The process of fine blanking with a V-shaped gripper

is of interest. For example when the shearing process takes place on an orthodox blanking tool, the deformation of a material occurs in the shearing zone and the breaking stress is attained before the punch has completely penetrated the material.

In fine blanking the breaking stress during shearing occurs only at the edges of the punch and die[2]. In the shearing zone the material is only deformed.

Words and Expressions

fine blanking			精密冲裁
cut edge			剪切刃
shaving	['ʃeɪvɪŋ]	v.	刨削,修整
producing hole			冲孔
cutting punch			冲裁凸模
burr	[bɜ:(r)]		[机]毛刺
bolster	['bəʊlstə(r)]	n.	垫板
clearance	['klɪərəns]	n.	间隙;(公差中的)公隙
orthodox	['ɔ:θədɒks]	adj.	传统的
shearing zone			剪切区

[1] The punch has a round edge and a small clearance and this is best used for blanks, but appears to give less satisfactory results when used for producing holes.

凸模有圆角边和小的间隙，最适合落料件，但是对于冲孔似乎并不能有较满意的结果。（句中"appear"译为"看来，似乎，出现"）

[2] In fine blanking the breaking stress during shearing occurs only at the edges of the punch and die.

精密冲裁中，剪切时破坏应力仅发生在凸凹刃口处。（短语"breaking stress"可译为"破坏应力"，"breaking strain"可译为"破坏应变"）

阅读材料

精密冲裁

精密冲裁是一种用于精确冲裁带平面产品的技术，它的剪切刃可比得上机加工的抛光。这种既快捷又方便的加工方法值得认真思考，特别是考虑到可省掉像刨削这样的操作，零件的数目也证实了冲裁工具的成本下降了。根据加工工件的类型可分为以下两种精密冲裁方法：

（1）凸模有圆角边和小的间隙，最适合落料件，但是对于冲孔似乎并不能有较满意的结果。

（2）这个方法使用一个沿外廓四周有 V 槽的 V 形夹板的冲裁凸模，在一个非圆形的凸模接触表面之前，夹具紧紧夹住落料件。

在第一种方法中，冲模的圆角半径由材料的硬度和厚度以及轮廓外形来决定。这个半径很重要，半径增加过大会导致变形和在落料件的下侧出现毛刺。因此，重要的是要找到一个最小半径，它能对零件产生好的影响。根据不同条件，这个半径可以在 0.3～2 mm 之间。

通常精密冲裁所需的加载只是传统冲裁所需加载的 10%。精密冲裁可以加工大半径和得到大的爆裂压力，因此，必须注意工具设计，这就意味着让冲模完全沉入的垫板必须有足够的面积。

为了确保消除凸模偏转，有必要固定凸模以便压住板块，否则，将导致失败，在设计中要注意这点。锋利的剪切刃也值得注意，一个明显的事实是凸模由高碳高铬钢制成。

工具设计中一个至关重要的问题是凸模和凹模间的间隙，它们比传统冲裁工具中的间隙都要小些。通常规定，总的间隙在 0.01～0.03 mm 之间为好，需要强调的是，这里所说的是总的间隙而不是每一个孔或落料件。

由于间隙小，将落料件从凸模中脱离的脱模力较高，脱出材料所需的加载比普通冲裁工具的高出 10% 左右。

第二种方法是在操作中用 V 形凸边围绕落料件并夹紧,这样凸边压进材料中。如图 16-4 所示为压力冲下时的情形,图中显示了压力板如何夹紧带钢、如何实现冲裁以及最终零件如何由下级卸料器排出。

有必要对普通冲裁和精密冲裁的不同作正确评价。例如,用传统的冲裁工具剪切加工时,材料的变形发生在剪切区,破坏应力发生在凸模完全穿透材料之前。

精密冲裁中,剪切时破坏应力仅发生在凸凹刃口处,在剪切区材料只产生了变形。

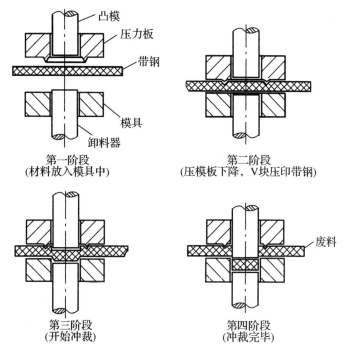

图 16-4　带 V 形夹板的精密冲裁过程

Unit 17

Gas-Assist Injection Molding

In the plastics processing arena, gas-assist injection molding is a relatively new process that is experiencing rapid growth[1]. Although it is a variation of injection molding, it is sometimes confused with blow molding. This is because both processes feature parts with hollow sections. One major difference between the processes is in the hollow section content of the part. Gas-assist molded parts have a much thicker wall surrounding a relatively small amount of hollow core. In general, gas-assist molded parts have less than 10% weight reduction in the hollow sections. Blow molding, on the other hand, can result in 80% or more hollow sections.

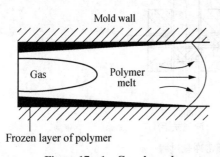

Figure 17 - 1   Gas channel

Gas-assist injection molding involves the injection of a short shot of resin into the cavity. When gas is introduced into the molten material, it takes the path of least resistance into areas of the part with low pressure and high temperature[2]. As the gas travels through the part, it cores out thick sections by displacing the molten material (Figure 17 - 1). This molten material fills out the rest of the part. After the filling is complete, the gas becomes the packing pressure, taking up the volumetric shrinkage of the material.

There are two basic types of gas-assist molding: constant volume and constant pressure. For the constant volume process, a cylinder of predetermined volume is pre-pressured prior to the gas injection. A piston pushes the gas out of the cylinder into the part. The pressure in the part depends on the ratio of part volume to the cylinder volume. Gas pressure, timing, and piston speed control the profile. For each injection cycle, the pressure must be built up prior to injection.

The constant pressure method functions with gas compressors that build up a reservoir of nitrogen to a predetermined pressure. This reservoir supplies a constant pressure to a series of valves. Pressure profiles are achieved by regulating the gas pressure to the valve and to the opening of each valve.

There are two primary options to implement gas-assist molding. These options differ in the location in which the gas is injected into the mold. Gas is injected either through the nozzle or directly into the mold cavity—either in the runner or directly into the part. The important difference is that the through-the-nozzle technique requires all gas channels to begin at the nozzle. When the gas is injected directly into the mold, the gas channel may be designed independent of the gate location, provided that the proper material fill pattern is achieved prior to gas injection.

Why is the level of interest in gas-assist injection molding so high? The primary reasons are that the "promised" advantages of this process are being delivered. Some of the potential benefits of gas-assist injection molding are:

(1) Reduced molded-in part stress

(2) Less warpage

(3) Reduced/eliminated sink

(4) Greater design freedom

(5) Increased parts integration possibilities

(6) Improved surface appearance

(7) Hollow sections provide easier part filling, longer material flow length, higher stiffness-to-weight ratio

(8) Reduced cycle time vs. solid sections

(9) Lower clamp tonnage requirements

(10) Lower injection pressures

(11) Reduced tooling costs as a result of replacing hot runner systems with gas channels

Gas-assist injection molding delivers the benefits. As a result, many companies are actively positioning new applications in this technology.

Some disadvantages must exist; otherwise, the process would have been implemented in a broader scale over a 25-year time period. The first issue is licensing. Table 17 - 1 lists the principal technology suppliers that require a license to be taken in order to employ their patented technologies. Equipment purchased for HELGA multi-nozzle process include the license.

Table 17 - 1 Gas-assist injection molding technology suppliers

Technology providers	System	License requirement
Battenfeld Airmould	Pressure regulation modules	No
CINPRES Ltd.	Multi-valve capability with integral pressure intensification	Yes
Epcon Gas Systems	Multiple valve systems with integral intensification; optional shot control methods	No

（**continued**）

Technology providers	System	License requirement
Ferromatik Milacron	Airpress Ⅲ	No
Gas Injection Limited	Pressure regulation modules；external gas-assist	Yes
GaIN Technologies	Through-the-nozzle；multiple valve systems with stand-alone nitrogen generation	Yes
HELGA(Hettinga)	Liquid injection molding	No
Incoe Corporation	External gas systems	No
Uniloy Milacron	Multi-nozzle injection molding based on multi-nozzle structural foam process	No
Nitrojection	Multi-valve systems with intensification capabilities	Yes

Another issue is additional cost. Even if a license is not required，process-specific equipment that ranges from ＄30，000 to ＄85，000 is required. These additional costs are incurred by the molder in order to have the capability to manufacture and sell components using gas-assist injection molding[3]. Depending on the license negotiated and the volume of parts manufactured，this fee can range from a few cents/unit to a significant additional cost.

Another source of added costs is the gas used in the process. Except for the HELGA process，nitrogen is the primary gas used due to its inert nature and plentiful supply. In the early stages of developing gas-assist injection molding capability，high-pressure cylinders were used as the nitrogen source. Why? At this point，volumes of nitrogen used are small，the tanks can be located close to the machine，and the cost is low.

As the quantity of parts molded using gas-assist injection molding increases，costs and safety issues associated with the nitrogen supply must by addressed. Options to high-pressure cylinders are liquid nitrogen in bulk cryogenic containers and on-site nitrogen generators.

Gas nozzle design and location are potential issues. With in-article and in-runner gas injection，the gas nozzle location is a critical tool design function. Polymer must cover the gas nozzle prior to gas introduction or blow-out will occur. Proper gas nozzle design and location are critical for optimum manufacturing productivity. Some nozzle designs will foul or plug during the gas injection or venting phase. This results in either maintenance or replacement，either of which will have a negative impact on productivity and cost.

Words and Expressions

arena	[ə'riːnə]	n.	竞技场,舞台
variation	[ˌveəri'eiʃn]	n.	变更,变化
cavity	['kævəti]	n.	模腔
molten	['məultən]	v.	熔化
volumetric	[vɒljʊ'metrɪk]	adj.	测定体积的
shrinkage	['ʃrɪnkɪdʒ]	n.	收缩
piston	['pɪstən]	n.	活塞
profile	['prəufaɪl]	n.	剖面,侧面,外形
compressor	[kəm'presə]	n.	压缩物,压缩机
reservoir	['rezəvwɑː]	n.	水库,蓄水池
nitrogen	['naɪtrədʒən]	n.	氮
regulate	['regjʊleɪt]	vt.	管制,控制,调节,校准
nozzle	['nɒzl]	n.	管口,喷嘴
deliver	[dɪ'lɪvər]	vt.	递送,释放
warpage	['wɔːpeɪdʒ]	v.	翘曲,扭曲,热变形
license	['laɪsəns]	vt.	许可,特许,认可
patent	['pætnt]	vt.	取得……的专利权
incur	[ɪn'kɜː]	v.	招致
negotiate	[nɪ'gəuʃieɪt]	v.	(与某人)商议,谈判,磋商
inert	[ɪ'nɜːt]	adj.	无活动的,惰性的,迟钝的
cryogenic	[ˌkraɪəʊ'dʒenɪk]	adj.	低温学的
generator	['dʒenəreɪtə]	n.	发电机,发生器
maintenance	['meɪntɪnəns]	n.	维护,保持
negative	['negətɪv]	adj.	否定的,消极的,负的
gas-assist injection molding			气辅注射成型
blow molding			吹塑成型
hollow section			中空截面
a short shot			欠料注射
fill out			充填
take up			解决
prior to			居先,在前
except for			除了……以外
associate with			联合
in bulk			散装,大批

blow-out   吹熄,爆裂

NOTES

〔1〕In the plastics processing arena，gas-assist injection molding is a relatively new process that is experiencing rapid growth.

在塑料加工成型的舞台上,气辅注射成型是一种新工艺,并正在迅速发展。(本句中 that 引导的从句 is experiencing rapid growth 的主语还是 gas-assist injection molding)

〔2〕When gas is introduced into the molten material，it takes the path of least resistance into areas of the part with low pressure and high temperature.

当气体被导入熔融的材料当中时,气体沿着阻力最小的方向流向制品的低压和高温区域。(本句中短语 of least resistance 修饰 path,而短语 with low pressure and high temperature 则修饰 areas of the part)

〔3〕These additional costs are incurred by the molder in order to have the capability to manufacture and sell components using gas-assist injection molding.

这些额外的费用是为了让制模工人具有使用气辅注射成型方法加工和出售零件的能力而产生的。(本句中短语 to manufacture and sell components using gas-assist injection molding 修饰 capability)

第 17 单元　气辅注射成型

在塑料加工成型的舞台上,气辅注射成型法是一种新工艺,并正在迅速发展。虽然它是注射成型的变异,但有时还和吹塑成型法相混淆。这是因为这两种工艺的特点都是中空截面。两者之间一个主要的区别是零件中空截面的容量不同。气辅成型的模制品具有比较厚的壁而围绕着相对少量的空心部分。通常,气辅成型的模制品在中空部分减少不少于10%的重量。而另一个方面,吹塑成型可以减少80%的重量或者产生更多的中空部分。

气辅注射成型法是向模腔内进行树脂的欠料注射。当气体被导入熔融的材料当中时,气体沿着阻力最小的方向流向制品的低压和高温区域。当气体在制品中流动时,它通过置换熔融物料而掏空厚壁截面(如图17-1所示)。这些置换出来的物料充填制品的其余部分。当填充过程完成以后,由气体继续提供保压压力,解决物料冷却过程中体积收缩的问题。

气辅注射成型有两种基本类型:定量成型和定压成型。定量成型工艺就是在气体注射之前把一个预先确定的圆筒形的材料进行预压。活塞再把气体从圆筒外面推进制品。制品内的压力依赖于制品体积和圆筒体积的比率。注气压力、注气时间和活塞速

率控制了制品的外形轮廓。每个定量注射周期，压力都必须在气体注射之前生成。

定压注射法是根据储存氮气的气体压缩机来预先确定压力的。这个氮气储存机为一系列的注射管提供了一个持续不变的压力。压力分布通过调节注射管的气压和注射管开口的气压来获得的。

要实现气辅注射成型有两个主要的方面。这两个方面由气体注入模具的位置不同而造成

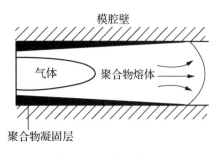

图 17-1　气体通道

的。气体可以经射出机的喷嘴进入模具，也可以直接进入，即可以经过流道进入制品，也可以直接进入制品。最重要的区别在于采用喷嘴这种工艺需要所有气体通道都始于射嘴。当气体直接注入模具中，假如在气体注射之前用适量的材料填充入模具，气体通道可以设计成独立的浇口位置。

为什么大家对气辅注射成型有如此高的兴趣呢？最主要的原因在于这种方法实现了所"许诺"的种种优点。气辅注射成型可能的一些优点是：

1. 降低制品内应力；

2. 减少制品翘曲变形；

3. 减少或消除缩水；

4. 提高设计自由；

5. 提高制品综合性能；

6. 改善制品表面质量；

7. 中空截面使制品充填更容易、材料流程更长、硬度/重量比更高；

8. 比实心截面更少的成型周期；

9. 更低的吨位机台锁模力；

10. 更低的注射压力；

11. 由于气道取代热流道而减少了模具成本。

气辅注射成型有以上这些益处。因此，很多公司都很积极地应用这项新工艺。

另外，气辅注射成型工艺必定存在一些缺点，否则，这项工艺会广泛应用超过25年的时间。首先最大的问题是许可证问题。表17-1列出了主要的技术提供者是否需要取得使用许可证才可以使用他们的专利技术。购买荷尔格公司或约翰控制公司的多喷嘴工艺设备，包含了使用许可证。

表 17-1　气辅注射成型的技术供应商

技术供应商	系统	需求许可证
巴顿菲尔公司	压力调节舱	无
辛普雷斯 股份有限公司	具有整体压力强化的多阀注射性能	有

续 表

技术供应商	系统	需求许可证
易可安气体系统公司	具有整体强化的多级管系统；任意注射控制方法	无
米拉克龙	气动压力 Ⅲ	无
气体注射有限公司	压力调节舱；外部气辅成型	有
耕毅科技	直通喷口；具有独立氮气发生器的多级管系统	有
荷尔格（海听格）	流体喷射法	无
英口公司	外部气辅系统	无
尤尼洛·米拉克公司	基于多喷口结构泡沫工艺的多喷口喷射法	无
注氮公司	具有强化性能的多阀注射系统	有

　　另外一个问题是额外的成本问题。即使不需要许可证，特殊的工艺设备也需要30 000美元到85 000美元。这些额外的费用是为了让制模工具有使用气辅注射成型方法加工和出售零件的能力而产生的。根据得到的授权和零件的产量，这个费用可大可小。

　　还有一个加入成本的是工艺中用到的气体。除了荷尔格工艺，氮气是用到的主要的气体，这归因于氮气的惰性和丰富的供应。在气辅注射成型的早期发展阶段，高压汽缸作为氮气的来源。为什么呢？在这时，氮气的使用量是很小的，氮气发生机可以装在注射机的附近，这样成本可以低一点。

　　当使用气辅注射成型法成型的零件质量不断提高时，必须要讨论成本、安全和氮气供应等因素。选择高压汽缸作为液态氮的大批低温容器和单边氮气发生器。

　　气体喷口的设计和定位是高成本可能存在的问题。在内模型气体注射和内流道气体注射中，气体喷嘴的位置设计是关键的设计。在气体注入之前，聚合物必须覆盖气体喷嘴，否则会发生工件吹裂的可能。适当的气体喷嘴设计和定位对获得最好的制造生产力是很重要的。一些喷嘴的设计将会在气体注射或气体排出时堵塞。这将导致需要维护或者替换喷嘴，否则会给生产力和成本带来负面的影响。

READING MATERIAL

Compression Molding

Compression molding is a method of molding in which the molding material, generally preheated, first placed in an open, heated mold cavity. The mold is closed with a top force or plug member; pressure is applied to force the material into contact with all mold areas, and heat and pressure are maintained until the molding material has cured[1]. Compression molding is a high-volume, high-pressure method suitable for molding complex, high-strength fiberglass reinforcements. The advantage of compression molding is its ability to mold large, fairly intricate parts.

Generally, in compression molding the molding material is plastic and the plastic material as powder preforms is placed into a heated steel mold cavity. Since the parting surface is in a horizontal plane, the upper half of the mold descends vertically. It closes the mold and applies heat and pressure for a predetermined period. A pressure of from 2 to 3 tons per square inch and a temperature at approximately 350 ℉ converts the plastic to a semi-liquid which flows to all parts of the mold cavity. Usually from 1 to 15 minutes is required for curing. Then the mold is opened and the part removed. If metal inserts are desired in the parts, they should be placed in the mold cavity on pins or in holes before the plastic is loaded. The preforms should be preheated before loading into the mold cavity to eliminate gases, improve flow, and decrease curing time. Thermosetting compression-molding compounds can be molded into articles of excellent rigidity and shape retention by supplying heat and pressure.

Apart from the molding material, the mold itself is of great importance. There are four types of molds for compression molding, namely, positive molds, landed positive molds, flash-type molds, and semi-positive molds. But the flash-type mold is widely used because it is comparatively easy to construct and it controls thickness and density within close limits[2].

A compression mold basically consists of an upper and a lower part. In normal cases the lower half is fitted to the table of the press and the upper half to the ram. Both mold halves are guided by hardened dowel. Asymmetrical parts cause large one-sided pressure loads to be exerted on the mold. They require compensation through special guides.

Compression molds nowadays are heated electrically exclusively. The mold is loaded with molding compound, by hand or with the aid of a filling device.

A construction drawing should be mandatory for every mold to be newly produced. Any new ideas concerning the mold, such as stability of the mold construction, optimum heating, aids to demolding and ejection, e.g., slides, split cavities, cores, etc., can be included in advance and given due consideration. The cost of such drawings will be more than justified as a rule by the ensuring efficient mold production and by fewer alterations and less finishing work on the completed mold. The more accurately details are incorporated in the design, the more finishing work is avoided, e.g., specification of the draft angle required and dimensional tolerances. Because alterations to compression molds are always very expensive, it is of particular importance that the detail drawings be completely clear.

Words and Expressions

predetermine	[ˌpriːdɪˈtɜːmɪn]	v.	预定,预先确定
semi-liquid			半流体
metal insert			金属嵌件
thermosetting	[ˈθɜːməʊˌsetɪŋ]	adj.	热硬化性的
mandatory	[ˈmændətəri]	adj.	命令的,强制的,托管的
namely	[ˈneɪmli]	adv.	即,也就是
positive mold			不溢式压缩模
landed positive mold			有肩不溢式压缩模
flash-type mold			溢出式压缩模
semi-positive mold			半溢式压缩模

NOTES

〔1〕The mold is closed with a top force or plug member；pressure is applied to force the material into contact with all mold areas，and heat and pressure are maintained until the molding material has cured.

模腔再通过上模来封闭,提供的压力迫使材料进入到模腔的每个角落,并且保持一定的热量和压力,直到成型材料固化。(本句中 contact with 是"接触"的意思,the material into contact with all mold areas,根据文意可译为"材料进入到模腔的每个角落")

〔2〕But the flash-type mold is widely used because it is comparatively easy to construct and it controls thickness and density within close limits.

但是溢出式压缩模在生产中使用最广泛,是因为它相对比较容易制造,并且能够控制塑件的厚度和致密度,使其接近要求。(本句中 easy to construct 是形容词＋不定式结构,可译为"容易制造")

阅读材料

压缩成型

压缩成型是一种成型方法,压缩成型的材料通常经过预热,然后放置在一个打开着的、热的模腔中。模腔通过上模来封闭,提供的压力迫使材料进入模腔的每个角落,并且保持一定的热量和压力,直到成型材料固化。压缩成型是一种高容量、高压力的成型方法,适合成型复杂的、高强度的玻璃纤维。压缩成型的优点在于可以成型大型的、复杂的零件。

通常压缩成型的材料是塑性材料。在压缩成型中,粉末状的塑料或预先成型过的塑料被放到加热的钢制模腔中。由于分型面是水平的,模具的上半部分垂直下来。它使模具闭合,然后在预塑阶段加热和加压。每平方英寸 2 到 3 吨的压力和大约 350 华氏度的温度使得塑料变为半流体流到模腔的每个部分。通常成型材料的固化时间需要 1 到 15 分钟。然后把模具打开,取出工件。如果零件中需要金属镶嵌件时,应该在塑料加载前将这些嵌件放入型腔的销孔中。预成型件应该在放入模具型腔之前预先加热,以消除材料中的气泡,改善材料的流动性,减少材料的固化时间。热硬性的压缩成型化合物在一定的保热和保压下可以成型为具有很好的硬度和形状的零件。

除了成型材料,压缩成型模具本身也非常重要。压缩模有四种类型的模具,即不溢式压缩模、有肩不溢式压缩模、溢出式压缩模和半溢式压缩模。但是溢出式压缩模在生产中使用最广泛,是因为它相对比较容易制造,并且能够控制塑件的厚度和致密度,使其接近要求。

一副压缩模通常由上、下两个部分组成。在正常情况下,模具的下半部分是安装在压力机的工作台上的,模具的上半部分是安装在滑块上的。上下两部分模具通过坚硬的销子导向。不对称的零件会对模具产生巨大的额外的偏心压力负荷。它们需要通过专门的导杆进行补偿。

现在,压缩成型通常有专门的电力加热。压缩模具通过手工或装料装置装载模塑料。

每个需要重新制造的模具都必须有一副结构图。任何关于模具的新的构思,比如模具的结构稳定性,最适宜的加热方法,如滑槽、分裂槽、型芯等预塑和注射的辅助结构,都可以包括在预期的考虑中。这些图纸花费的成本在整副模具确保有效生产、更少修模以及更少修整工件等方面被证明是值得的。设计中涉及的细节越精确,就越能避免修整工件。例如,所需的拔模斜度和尺寸公差的说明书。因为更换压缩模很昂贵,所以十分清楚有条理的细节图尤其重要。

Unit 18

Electrical Discharge Machining

Electrical discharge machining(EDM) is the process of electrical discharging between tool and work which result in the material is eroded. It uses an electric spark to erode the unwanted material from the workpiece, which takes a shape opposite to that of the electrode(shown as Figure 18 – 1)[1]. Due to the electrical discharge short-lived, but very high, temperature rises, metal particles at the point are molten, partially vaporized and removed from the melt by mechanical and electromagnetic forces. The electrode and the workpiece are both submerged in a dielectric fluid, which is generally a light lubricating oil. The dielectric fluid, which is constantly being circulated, washes the eroded material away and simultaneously acts as a coolant.

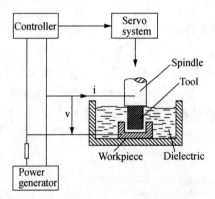

Figure 18 – 1　Basic elements of an electro-discharge system

There is no contact between the electrode and the work and customarily maintains a gap of about 0.0005 to 0.001 in(0.01 to 0.02 mm). The electrode, made to the shape of the cavity required, is made from electrically conductive material, usually carbon[2]. Owing to some erosion of the electrode in EDM, there must be allowed for in the tool design to ensure accuracy of machining. In EDM, there is no fixed tool feed, but the gap size must be maintained in accordance with the rare of metal removal and the conditions existing within the gap[3].

Compare with the conventional machining processes, the electrical discharge machining has many advantages. Taking one with another, the merits are as follows. Firstly, electrical discharge machining can process any material that is electrically conductive. And it is especially valuable for cemented carbides that are difficult to cut by conventional means. Moreover it can be easily machining thin or fragile sections without deforming and at the same time secondary finishing operations are eliminated for many types of workpiece[4]. Secondly, there is no stress in the work material because the electrode never comes in contact with the workpiece in EDM. And the workpiece can be machined in a hardened state, consequently overcoming the deformation which caused by the hardening process. Thirdly, intricate shapes of die-sinking, which impossible to produce by conventional means, can be cut out of a solid work with relative ease by electrical discharge machining. And there is no burr on the workpiece which produced by EDM, carry out better dies and molds produced at lower costs and high efficiency. Finally, only one person can operate several electrical discharge machining machines at one time and economize costs.

There are many machining methods that the process principles are similar to the spark erosion machining, such as traveling-wire EDM, electron-beam machining (EBM) and laser cutting machining and so on.

1. Traveling-Wire EDM

As with all electrical discharge machining, the actual metal removal is the result of a spark discharge jumping the gap from a tool through a dielectric to the workpiece. Then intense heat is created in the localized area of the spark impact, the metal melts and a small particle of molten metal is expelled from the surface of the workpiece. In time the workpiece is simply eroded away. The traveling-wire method employs a reel of copper wire that is slowly fed past the workpiece and the wire is used only once. That is to say, the cutting tool is a thin copper wire, which enters the work during cutting without physical contact(shown as Figure 18 – 2).

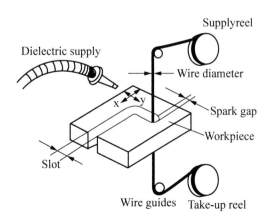

Figure 18 – 2　Traveling-wire electrical discharge machining

This suffers wear as a result of the action of spark erosion, and for this reason fresh wire is constantly supplied. In jumping from the wire to the workpiece, the sparks erode away a clearly defined path, which can be very closely controlled with NC. Hardened tool steels and carbides are effectively machined with this method.

2. Electron-Beam Machining

Electron-Beam Machining(EBM) is a machining process where high-velocity electrons are directed toward a workpiece, creating heat and vaporizing the material. The fundamental of electron-beam machining is as follows. EBM machines utilize voltages in the range of 50 to 200 kV to accelerate electrons to 200,000 km/s and then electromagnetic lenses are used to direct the electron beam, by means of deflection, into a vacuum. When the electrons strike the workpiece, electrons' kinetic energy change into heat energy, and produce high temperature which can melt and vapor any material. Because work must be done in a vacuum, EBM is best suited for small parts.

3. Laser Cutting Machining

Laser cutting uses the erosion effect of high-energy light beams. As in the case of electron-beam cutting, the work material is vaporized at the point of impact. Within a few milliseconds, a channel is cut into the work material. The vapor pressure forces the molten metal in the immediate vicinity out of the channel.

Words and Expressions

discharge	[dɪsˈtʃɑːdʒ]	n.	放电,卸货,流出
EDM = electrical discharge machining			电火花加工
erode	[ɪˈrəʊd]	v.	腐蚀,侵蚀,使变化,受腐蚀
electrode	[ɪˈlektrəʊd]	n.	电极
vaporize	[ˈveɪpəraɪz]	v.	(使)蒸发,(使)汽化
submerge	[səbˈmɜːdʒ]	v.	浸没,淹没
dielectric	[ˌdaɪɪˈlektrɪk]	adj.	非传导性的
coolant	[ˈkuːlənt]	n.	冷冻剂,冷却液,散热剂
conductive	[kənˈdʌktɪv]	adj.	传导的
conventional	[kənˈvenʃnl]	adj.	惯例的,常规的,习俗的
overcome	[ˌəʊvəˈkʌm]	vt.	战胜,克服,胜过,征服
		vi.	得胜
intricate	[ˈɪntrɪkət]	adj.	复杂的,错综的,难以理解的
economize	[ɪˈkɒnəmaɪz]	v.	节约,节省,有效地利用
gap	[ɡæp]	n.	缺口,裂口,间隙,缝隙,差距

impact	[ˈɪmpækt]	n.	碰撞,冲击,影响,效果
expel	[ɪkˈspel]	v.	卷……于轴上,使旋转,旋转
wear	[weə]	v.	(～out)磨损,用旧
fresh	[freʃ]	adj.	新鲜的,无经验的,生的
in accordance with			与……一致
compare with			与……比较
cemented carbides			硬质合金,烧结碳化物
die-sinking			加工模腔,开模
electro-discharge cutting machine			电火花切割机
electrical discharge forming			电火花成型机
traveling-wire EDM machine			线电极电火花加工机床
EDM grinders			电火花加工磨床
cavity sinking EDM machines			型腔电火花加工机床
wire cut electric discharge machine			电火花线切割机

NOTES

〔1〕 It uses an electric spark to erode the unwanted material from the workpiece, which takes a shape opposite to that of the electrode(shown as Figure 18-1).

它使用电火花浸蚀工件上多余的金属,工件在切削后的形状与电极相反(如图 18-1 所示)。(本句主语 It 指电火花加工,由 which 引导的从句修饰 workpiece,从句中的形容词短语 opposite to that of the electrode 作定语修饰 shape,其中 that 指代 shape)

〔2〕 The electrode, made to the shape of the cavity required, is made from electrically conductive material, usually carbon.

电极用导电材料制造,通常是碳,且电极的形状与所需型腔匹配。(本句中过去分词短语 made to the shape of the cavity required 作定语,用来修饰句子主语 the electrode)

〔3〕 In EDM, there is no fixed tool feed, but the gap size must be maintained in accordance with the rare of metal removal and the conditions existing within the gap.

在电火花加工中,刀具进给量不是固定的,加工间隙的大小依据金属的切削速率和该加工间隙的工作状态来确定。(本句句尾的分词短语 existing within the gap 作为 the conditions 的后置定语,可译为"加工间隙的工作状态")

〔4〕 Moreover it can be easily machining thin or fragile sections without deforming and at the same time secondary finishing operations are eliminated for many types of workpiece.

电火花加工对薄而脆的工件可以很轻易地进行无变形加工,同时对加工处理的许多类型的工件,免除二次精加工。

第18单元　电火花加工

电火花加工是在刀具和工件之间产生放电使材料被腐蚀的一种加工工艺。它使用电火花浸蚀工件上多余的金属,工件在切削后的形状与电极相反(如图18-1所示)。由于放电过程很短暂,温度升得很高,放电处的金属微粒被熔化,熔化的材料一部分蒸发,其余的由机械力和电磁力从未熔化的材料上去除。工件和电极都浸在绝缘的液体里,这种液体通常是轻润滑油。这种不断循环的绝缘的液体把已腐蚀下来的材料带走,同时还起着冷却液的作用。

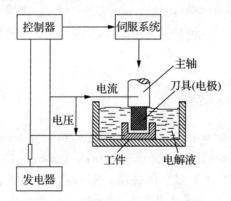

图18-1　电火花加工的基本元素

在电火花加工中,电极和工件并不接触,通常保持0.000 5～0.001英寸(0.01～0.02毫米)的间隙。电极用导电材料制造,通常是碳,且电极的形状与所需型腔匹配。由于在电火花加工中对电极有一定的腐蚀,在刀具设计中必须有一定的余量以保证加工精度。在电火花加工中,刀具进给量不是固定的,加工间隙的大小依据金属的切削速率和该加工间隙的工作状态来确定。

和传统的机加工工艺相比,电火花加工有很多的优点。总的来说,有以下几个优点。第一,电火花加工方法可以加工任何导电的材料。对用传统的方法难以切削的硬质合金,电火花加工特别有价值。电火花加工对薄而脆的工件可以很轻易地进行无变形加工。同时对加工处理的许多类型的工件,免除二次精加工。第二,电火花加工中电极与工件未接触,所以工件不会产生应力,并且工件可在淬火状态下加工,克服了由淬火引起的变形问题。第三,电火花加工能相对容易地在实心毛坯上加工出传统方法不能加工出来的复杂形状的型腔,加工出的零件还无毛刺,能低成本高效率地加工出比较好的模具。最后,只一个人就可以操作好几台电火花加工机床,节约了成本。

采用与电火花加工原理相似的加工方法有很多,如:线电极电火花加工、电子束加工以及激光切割法等等。

1. 线电级电火花加工

线电极电火花加工和所有的电火花加工一样,电火花跳过电极与工件间的不导电的液体间隙,从而去除金属。在火花冲击的局部地区产生大量的热量,金属熔化,并从工件表面喷出熔化的金属的小粒子。工件即被腐蚀掉了。线电极方法就是使用一卷铜线,缓慢地经过工件进给,这些线只使用一次。也就是说,切割刀具是极细的铜丝,在切割过程中进入工件而与工件无物理接触(如图18-2所示)。

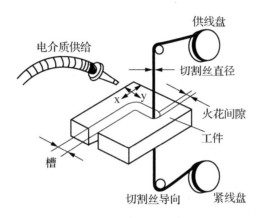

图 18-2　线电极电火花加工

电火花的产生使得电极丝有磨损,因此,需要不断地提供新的切割丝。电火花从电极丝跳跃到工件,经过设定好的路线腐蚀材料,这些路线可以由数字计算机精确控制。硬质工具钢和碳化物都可以用线电极电火花,加工方法来加工。

2. 电子束加工

电子束加工(简称EBM)是利用高速的电子束冲击工件时所产生的热能使材料熔化、气化的一种加工方法。电子束加工的基本原理如下:电子束机床用50～200千伏的高电压把电子加速到200 000千米/秒,然后在真空中,通过电磁透镜会聚成一束高功率密度电子束。当电子束冲击到工件时,电子束的动能转变成为热能,产生出极高的温度,足以使任何材料瞬时熔化、气化。由于加工必须在真空中进行,因此,电子束加工比较适合加工小型零件。

3. 激光切割加工

激光切割是采用高能量的光束的侵蚀作用来加工材料的。和电子束切割一样,激光切割加工的材料在光束冲击材料的时候被气化。在极短的时间内,在工件材料上切割出一个切缝。再用高气压把熔化的金属从切缝中吹走。

READING MATERIAL

Electrochemical Machining

When applying the electrochemical machining(ECM) process, the electrochemical

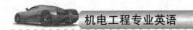

reaction dissolves metal from a workpiece into an electrolyte solution. A current is passed through an electrolyte between a positively charged anodic workpiece and a negatively charged cathode tool, which causes the metal to be removed ahead of the tool as the tool is fed to the workpiece. And then the workpiece dissolves whilst the tool is not affected. The principle underlying electrochemical machining rests on an exchange of charge and material between a positively charged anodic workpiece and a negatively charged cathode tool in an electrolyte.

The current in the electrolyte causes chemical reaction which dissolves the metal from the workpiece. And the current is dependent on the gap between the workpiece and the tool. The electrode and the workpiece locate within 0.002 to 0.003 in, which allows electrochemical machining to take place without physical contact. That is to say, the working gap maintained between the electrode and workpiece allows machining to take place without physical contact[1]. The electrolyte solution is a controlled flowing stream which carries the current and the electrolyte also serves as a coolant. The supply voltage is dependent on the area of erosion and the conductivity of the electrolyte.

The electrode for ECM is not a simple bar of metal. When producing internal forms, a design problem arises in relation to the shape and size of the electrode. For example, a cylindrical bore is to be sunk, a simple cylindrical electrode results in a constantly increasing gap size and the current density decreases in proportion. But with a precision insulated electrode which has been made to a specific shape and exact size, the offending excessive erosion of cylindrical sides are suppressed.

It is necessary for electrochemical machining to have high electrolyte pressures so that there is an adequate flow for effective cooling and removal of the eroded material because the small gap sizes are used[2]. As the electrolyte solution consists of a corrosive salt solution, all electrochemical machine components likely to come into contact with it must be corrosion-proof.

Electrochemical machines have two basic structure types which are the open "C" structure for small-to-medium sized work and the "O" structure for having larger work areas. And usually the tool feed direction of electrochemical machining is vertical. Mostly the tool advance is electro-hydraulic, but electric drives are also used.

There are several advantages of using electrochemical machining. First, it can machine metal of any hardness and compete with drilling especially if the workpiece exceeds 42 Rockwell C hardness. Second, it can machine intricate forms and produce easily which difficult to machine by other processes. Third, there is no heat created during the machining process and there is no workpiece distortion. Forth, owing to the tool never touches the workpiece, tool wear is insignificant and thin fragile sections can be machined without distortion.

Words and Expressions

electrochemical	[ɪˌlektrəʊˈkemɪkl]	*adj.*	电化学的
ECM = electrochemical machining			电化学加工
dissolve	[dɪˈzɒlv]	*v.*	溶解,解散
arise	[əˈraɪz]	*vi.*	出现,发生,起因于
positively charged			接正极的
anodic	[əˈnɒdɪk]	*adj.*	阳极的
negatively charged			接负极的
cathode	[ˈkæθəʊd]	*n.*	阴极
exchange	[ɪksˈtʃeɪndʒ]	*n.*	调换,兑换,交流,交易
corrosion-proof			防腐

NOTES

〔1〕That is to say, the working gap maintained between the electrode and workpiece allows machining to take place without physical contact.

也就是说,电极和工件之间保持的加工间隙,既避免了物理接触,又能完成加工。(本句中 That is to say 可译为"也就是说""更确切地说"。分词 working 作 gap 的前置定语,分词短语 maintained between the electrode and workpiece 作 gap 的后置定语)

〔2〕It is necessary for electrochemical machining to have high electrolyte pressures so that there is an adequate flow for effective cooling and removal of the eroded material because the small gap sizes are used.

由于电解加工中的间隙很小,电解加工必须要有高的电解压力使电解液具有足够的流动力以有效冷却和去除已被腐蚀掉的材料。(本句中 It is necessary for electrochemical machining to have 是 it is necessary for sb. to do sth.句型。结果状语从句 so that there is an adequate flow for effective cooling and removal of the eroded material 用来修饰 high electrolyte pressures)

阅读材料

电化学加工

当采用电化学加工工艺时,电解反应把金属从工件上溶解到电解液中。电流通过接正极的阳极工件和接负极的阴极刀具之间的电解液,使得金属在刀具向工件进给时在刀具的前面去除。从而工件被溶解而刀具不受影响。电解加工的基本原理是在电解液中接正极的阳极工件和接负极的阴极刀具之间的电荷和材料的交换。

在电解液中的电流产生了能使金属从工件上溶解的化学反应。电流依赖于工件和刀具之间的间隙。电极和工件之间相距 0.002 到 0.003 英寸,这个距离可以使得进行电化学加工而电极和工件没有物理接触。也就是说,电极和工件之间保持的加工间隙,既避免了物理接触,又能完成加工。电解液是一股受控制的流动的液体,可以传递电流,电解液还起着冷却液的作用。所需提供的电压则依赖于腐蚀面积和电解液的传导率。

电化学加工的电极不是一个简单的金属棒。当加工内加工面时,就产生了电极的形状和尺寸的设计问题。例如,要加工一个圆柱形的孔,一个简单的圆柱形电极会导致间隙尺寸不断增加而电流密度则成比例减小。但是如果采用一个已被加工成具有精确外形和精密尺寸的具有一定精度的绝缘电极,则圆柱形电极外侧的额外的材料腐蚀就会避免。

由于电解加工中的间隙很小,因此,电解加工必须要有高的电解压力才会使电解液具有足够的流动力以有效冷却和去除已被腐蚀掉的材料。由于电解液是由腐蚀盐溶液组成的,因此,所有可能和电解液接触的电化学加工机床的元件必须是防腐蚀的。

电化学机床具有两种基本结构类型,分别是用于中小型尺寸的工件加工的"C"型结构和具有大的工作空间的"O"型结构。通常电化学加工的刀具的进给方向是垂直的。大多数的刀具进给是用电液装置来实现的,但是也可以用电装置来实现。

采用电化学加工有很多优点。第一,它可以加工任何硬度的金属,并且尤其在洛氏硬度超过 42HRC 时,采用电化学加工比钻削加工更具有竞争力。第二,它可以加工复杂形状的工件,并且可以加工其他工艺很难加工的工件。第三,在电化学加工过程中不产生热量,因此,也没有工件的变形。第四,由于刀具与工件没有接触,因此,刀具磨损非常小,并且加工细小易脆的部分时不易变形。

Unit 19

Membrane Stresses in Shells of Revolution

扫码可见本项目
☞ 参考资料

A shell of revolution is the form swept out by a line or curve rotated about an axis(A solid of revolution is formed by rotating an area about an axis[1]). Most process vessels are made up from shells of revolution: cylindrical and conical sections; and hemispherical, ellipsoidal and torispherical heads. Typical vessel shapes are shown in Figure 19 - 1.

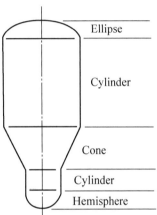

Figure 19 - 1   Typical vessel shapes

The walls of thin vessels can be considered to be "membranes"; supporting loads without significant bending or shear stresses; similar to the walls of a balloon.

The analysis of the membrane stresses induced in shells of revolution by internal pressure gives a basis for determining the minimum wall thickness required for vessel shells[2]. The actual thickness required will also depend on the stresses arising from the other loads to which the vessel is subjected.

Consider the shell of revolution under a loading that is rotationally symmetric; that is, the load per unit area (pressure) on the shell is constant round the circumference, but not necessarily the same from top to bottom.

Let P = pressure

   t = thickness of shell

   σ_1 = the meridional(longitudinal)stress, the stress acting along a meridian

   σ_2 = the circumferential or tangential stress, the stress acting along parallel circles(often called the hoop stress)

   r_1 = the meridional radius of curvature

   r_2 = circumferential radius of curvature

Note: the vessel has a double curvature; the values of r_1 and r_2 are determined by the shape.

Considering the forces acting on the element, and equating these forces and simplifying, gives:

$$\frac{\sigma_1}{r_1}+\frac{\sigma_2}{r_2}=\frac{P}{t} \qquad (19-1)$$

$$\sigma_1=\frac{Pr_2}{2t} \qquad (19-2)$$

Equations(19 - 1) and (19 - 2) are completely general for any shell of revolution.

1. Cylinder

A cylinder[Figure 19 - 2(a)] is swept out by the rotation of a line parallel to the axis of revolution, so:

$$r_1=\infty, r_2=D/2$$

where D is the cylinder diameter.

Substitution in equations (19 - 1) and (19 - 2) gives:

$$\sigma_2=\frac{PD}{2t}, \sigma_1=\frac{PD}{4t} \qquad (19-3)$$

2. Sphere

A sphere[Figure 19 - 2(b)] is swept out by a semicircle rotated about the axis of diameter, here:

$$r_1=r_2=D/2$$

hence:

$$\sigma_1=\sigma_2=\frac{PD}{4t} \qquad (19-4)$$

3. Cone

A cone[Figure 19 - 2(c)] is swept out by a straight line inclined at an angle α to the axis.

$$r_1=\infty, r_2=r/\cos\alpha$$

Substitution in equations(19 - 1) and (19 - 2) gives:

$$\sigma_1=\frac{Pr}{2t\,\cos\alpha}, \sigma_2=\frac{Pr}{t\,\cos\alpha} \qquad (19-5)$$

The maximum values will occur at

$$r=D_2/2 .$$

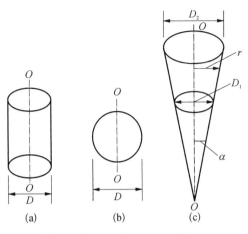

Figure 19 - 2   Typical revolutions

Words and Expressions

membrane	[ˈmembreɪn]	n.	薄膜,隔膜,隔板
revolution	[ˌrevəˈluːʃn]	n.	回转,旋转;运行,公转
cylindrical	[səˈlɪndrɪkl]	adj.	圆柱(体形)的,(圆)筒形的
conical	[ˈkɒnɪkl]	adj.	圆锥的,圆锥形的,圆锥体的
hemispherical	[ˌhemɪˈsferɪkl]	adj.	半球形的,半球状的
ellipsoidal	[ɪˈlɪpsɒɪdl]	adj.	椭圆形(的),椭圆球(的),椭圆体的
torispherical	[tɒrɪˈsferɪkl]	adj.	准球形的
induce	[ɪnˈdjuːs]	vt.	引起,导致
symmetric	[sɪˈmetrɪk]	adj.	对称的,均衡的,平均的
longitudinal	[ˌlɒndʒɪˈtjuːdɪnl]	adj.	经线的,纵向的,轴向的
circumferential	[səˌkʌmfəˈrenʃəl]	adj.	圆周的,环形的
rotationally symmetric			轴对称的
curvature	[ˈkɜːvətʃə]	n.	弯曲;弯曲部分;曲率,曲度
resist	[rɪˈzɪst]	vt.	不受(某事物)损害[影响];抗,耐
equate	[ɪˈkweɪt]	vt.	立方程式,使……等于
meridional	[məˈrɪdɪənl]	adj.	子午线的,经向的
substitution	[ˌsʌbstɪˈtjuːʃn]	n.	代替,取代作用;代入法,代入

NOTES

[1] A shell of revolution is the form swept out by a line or curve rotated about

an axis(a solid of revolution is formed by rotating an area about an axis).

回转壳体是由一条直线或曲线绕着一根轴旋转形成的曲面(回转体是由一个面绕一轴线旋转而成的)。(句中 sweep out 可译为"扫过")

[2] The analysis of the membrane stresses induced in shells of revolution by internal pressure gives a basis for determining the minimum wall thickness required for vessel shells.

对由内压引起的回转壳体的薄膜应力的分析为确定容器壳体所需的最小壁厚奠定了基础。(该句是一个较长的简单句,分析这样的句子应首先找出句子的主干部分,即 The analysis gives a basis,其他部分为修饰语。induced in shells of revolution by internal pressure 为过去分词短语,作 stresses 的后置定语。同样,required 也为过去分词,作 thickness 的定语)

第 19 单元　回转壳体的薄膜应力

回转壳是由一条直线或曲线绕着一根轴旋转形成的曲面(回转体是由一个面绕一根轴旋转而成)。多数过程容器由回转壳体组成:圆筒形和圆锥形部分;加上半球形、椭球形和准球形封头典型的容器形状,如图所示(19-1):

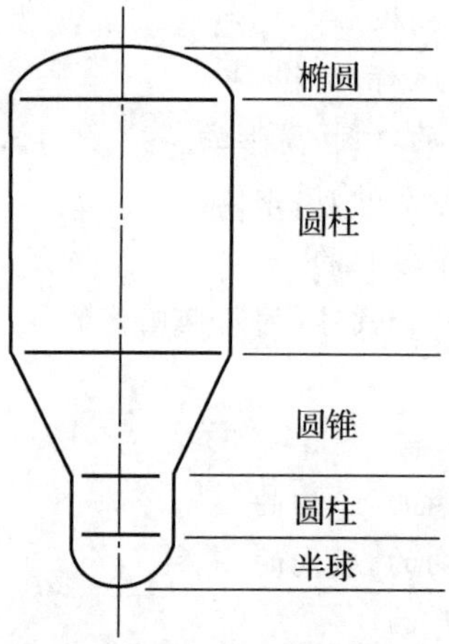

容器的器壁可称为"薄膜",所受弯曲应力和剪应力意义不大,类似于气球的球壁。

对由内压引起的回转壳体的薄膜应力分析为确定容器壳体所需的最小壁厚奠定了基础。容器实际所需厚度同时也取决于容器承受的其他载荷所引起的应力。

考察在轴对称载荷作用下的回转壳体,即壳体中沿环向每单位面积上的载荷(压力)是不变的,但是从壳体顶部到底部的载荷无须相同。

令 P = 压力

t = 壳体厚度

σ_1 = 经向(轴向)应力,即沿经向的应力

σ_2 = 周向或切向应力,即沿平行圆方向的应力(常称作环向应力)

r_1 = 经向曲率半径

r_2 = 周向曲率半径

图 19-1　典型容器形状

椭圆

圆柱

圆锥

圆柱

半球

注:容器有两个曲率半径,r_1 和 r_2 的数值大小由形状而定。

考察作用在单元体上的力,然后通过计算和简化,给出:

$$\frac{\sigma_1}{r_1} + \frac{\sigma_2}{r_2} = \frac{P}{t} \qquad (19-1)$$

$$\sigma_1 = \frac{Pr_2}{2t} \tag{19-2}$$

式(19-1)和(19-2)是任意回转壳体的普遍方程。

1. 圆筒壳体

圆筒壳体[如图19-2(a)所示]是由一根直线围绕与其平行的回转体中心轴线旋转而成的,所以:

$$r_1 = \infty, r_2 = D/2$$

在这里,D 为圆筒直径。

代入式(19-1)和(19-2)给出:

$$\sigma_2 = \frac{PD}{2t}, \sigma_1 = \frac{PD}{4t} \tag{19-3}$$

2. 球形壳体

球形壳体[如图19-2(b)所示]是由一根半圆周线绕其直径轴旋转而成的,这里:

$$r_1 = r_2 = D/2$$

因此:

$$\sigma_1 = \sigma_2 = \frac{PD}{4t} \tag{19-4}$$

3. 圆锥壳体

圆锥壳体[如图19-2(c)所示]是由一根直线绕着与之倾角为 α 的轴旋转而成的。

$$r_1 = \infty, r_2 = r/\cos\alpha$$

代入式(19-1)和(19-2)给出:

$$\sigma_2 = \frac{Pr}{t\cos\alpha}, \sigma_1 = \frac{Pr}{2t\cos\alpha} \tag{19-5}$$

应力最大值发生在

$$r = D_2/2 \text{。}$$

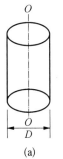
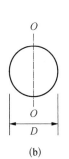
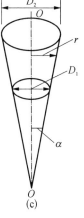

(a)　　　　(b)　　　　(c)

图 19-2　典型回转壳体

READING MATERIAL

Pressure Vessels and Their Components

Typical components of pressure vessels are described below.

1. Cylindrical Shell

Cylindrical shell is very frequently used for constructing pressure vessels in the petrochemical industry. It is easy to fabricate and install，and economical to maintain. The required thickness is generally controlled by internal pressure，although in some instances applied loads and external pressure have control. Other factors such as thermal stress and discontinuity forces may also influence the required thickness.

2. Formed Heads

A large variety of end closures and transition sections are available to the design engineer. Using one configuration versus another depends on many factors such as method of forming，material cost，and space restriction. Some frequently used formed heads are：

(1) Flanged heads. These heads are normally found in vessels operating at low pressure such as gasoline tanks，and boilers. They are also used in high pressure applications where the diameter is small. Various details for their design and construction are given by the ASME Code，Ⅷ-1.

(2) Hemispherical heads. Generally，the required thickness of hemispherical heads due to a given temperature and pressure is one-half that of cylindrical shells with equivalent diameter and material. Hemispherical heads are very economical when constructed of expensive alloys such as nickel and titanium—either solid or clad. In carbon steel，however，they are not as economical as flanged and dished heads because of the high cost of fabrication. Hemispherical heads are normally fabricated from segmental "gore" sections or by spinning or pressing. Because hemispherical heads are thinner than cylindrical shells to which they are attached，the transition area between the heads and shell must be contoured so as to minimize the effect of discontinuity stress[1].

(3) Elliptical and torispherical(flanged and dished) heads. These heads are very popular in pressure vessels. Their thickness is usually the same as the cylinder to which they are attached. This reduces considerably the weld build-up. Thus，because the required thickness in areas away from the knuckle region is less than the furnished thickness，the excess can be advantageously used in reinforcing nozzles in these areas[2]. Many mills can furnish such heads in various diameters and thickness that are competitive in price.

(4) Conical and toriconical heads. These heads are used in hoppers and towers as bottomed closures or as transition sections between cylinders with different diameters. The cone-to-cylinder junction must be considered as a part of the cone design due to the high unbalanced forces at the junction. Because of these high forces，the ASME Code，Ⅷ-1，limits the apex angle to a maximum of 30°when the

cone is subjected to internal pressure. Above $30°$ a discontinuity analysis is done or a toriconical head used to avoid the unbalanced forces at the junction.

3. Blind Flanges, Cover Plates, and Flanges

One of the more common types of closures for pressure vessels is the unstayed flat head or cover. This may be either integrally formed with the shell or welded to the shell，or it may be attached by bolts or some quick-opening device. It may be circular，square，rectangular，or some other shape. Those circular flat heads that are bolted into place utilizing a gasket are called blind flanges. Usually，the blind flange is bolted to a vessel flange with a gasket between two flanges. Although flat heads or blind flanges may be either circular or noncircular，they usually have uniform thickness.

4. Openings and Nozzles

All process vessels require openings to get the contents in and out. For some vessels，where the contents may be large or some of the internal parts may need frequent changing，access is made through large openings in which the entire head or a section of the shell is removed. However，for most process vessels，the contents enter and exit through openings in the heads and shell to which nozzles and piping are attached. In addition to these openings others may be required，such as those for personnel entering the vessel through a manway opening. Other openings may be necessary for inspecting the vessel from outside through a handhole opening，and still others may be required for cleaning or draining the vessel. These openings do not always have a nozzle located at the opening. Sometimes the closure may be a manway cover or handhole cover that is either directly welded or attached to the vessel by bolts.

5. Supports

Most vertical vessel are supported by skirts. Skirts are economical because they generally transfer the loads from the vessel by shear action. They also transfer the loads to the foundation through anchor bolts and bearing plates. Leg-supported vessels are normally lightweight and the legs provide easy access to the bottom of the vessel. An economic design is that the legs attach directly to the vessel and the loads are transferred by shear action. Horizontal vessels are normally supported by saddles. Stiffening rings may be required if the shell is too thin to transfer the loads to the saddles. The problem of the thermal expansion must be considered.

Words and Expressions

vessel　　　　　['vesl]　　　　　　　*n*.　　　容器，器皿；槽；罐

discontinuity	[ˌdɪsˌkɒntɪˈnjuːəti]	n.	断绝,不连续,中断,不均匀
transition	[trænˈzɪʃn]	n.	过渡;转变;变迁;转折
flange	[flændʒ]	vt.	折边,安装凸缘
gasoline	[ˈɡæsəliːn]	n.	汽油
clad	[klæd]	adj.	覆盖的
nickel	[ˈnɪkəl]	n.	[化]镍
titanium	[tɪˈteɪniəm]	n.	钛
dish	[dɪʃ]	vt.	盛于碟盘中;使成碟形
segmental	[seɡˈmentl]	adj.	部分的,分片的
gore	[ɡɔː]	n.	三角地带
torispherical	[ˌtɒrɪˈsferɪkl]	adj.	准球形的
contour	[ˈkɒntʊə]	vt.	使与轮廓相符
knuckle	[ˈnʌkl]	n.	转向节,万向接头
reinforce	[ˌriːɪnˈfɔːs]	vt.	增强;加强;补充
toriconical	[ˌtɒrɪˈkɒnɪkl]	adj.	准锥形的
hopper	[ˈhɒpə]	n.	加料斗,漏斗
unstayed	[ʌnˈsteɪd]	adj.	未固定的,未加固的,无支撑的
manway	[ˈmænweɪ]	n.	人孔,(煤矿等内的)人行巷道
handhole	[ˈhændhəʊl]	n.	手孔
skirt	[skɜːt]	n.	裙座
anchor	[ˈæŋkə]	n.	锚,钩子
anchor bolt			地脚螺栓
saddle	[ˈsædl]	n.	车座,鞍座

NOTES

〔1〕Because hemispherical heads are thinner than cylindrical shells to which they are attached, the transition area between the heads and shell must be contoured so as to minimize the effect of discontinuity stress.

由于半球形封头比与之相接的圆筒形壳体要薄,因此,封头与壳体之间的过渡区域必须逐渐变化,以减小不连续应力的影响。

〔2〕Thus, because the required thickness in areas away from the knuckle region is less than the furnished thickness, the excess can be advantageously used in reinforcing nozzles in these areas.

因此,由于折边区以外的区域所需的厚度小于封头的实际厚度,故,多余的部分就可以用于这些区域内的接管补强。

■ 阅读材料

压力容器及其零部件

压力容器典型零部件如下所述:

1. 圆筒形壳体

圆筒形壳体是石油化学工业中构成压力容器常用的一种形式。它易于制造和安装,并且维护费用经济实惠。尽管某些情况下,壳体所需厚度受施加的载荷和外压影响,但其厚度往往是由内压控制的。其他因素如热应力和不连续应力也影响所需厚度。

2. 成形封头

对设计工程师而言,可选取各种端盖部封闭形式和过渡部件。使用某种形状而不用另一种取决于很多因素。例如,成形的方法、材料成本和空间限制。一些常用的成形封头:

(1) 凸缘封头。这些封头常用于低压操作的容器上。例如汽油槽和锅炉。它们也用于小直径的高压设备上。ASME 规范的第八篇第一分篇给出了这类封头的各种设计和结构的细节。

(2) 半球形封头。通常,由给定温度和压力决定的半球形封头所需厚度是相同直径、相同材料的圆筒封头厚度的一半。当使用如镍和钛这种昂贵合金材料制成的封头——无论是整体还是覆层,半球形封头的价格都是非常经济的。而用碳钢的话,半球形封头不如凸缘封头和碟形封头经济,因为碳钢半球形封头制造成本高。半球形封头通常由"三角形带"分片制成,或由旋压或冲压而成。由于半球形封头比与之相接的圆筒形壳体要薄,因此,封头与壳体之间的过渡区域必须逐渐变化,以减小不连续应力的影响。

(3) 椭圆形和准椭圆形(凸缘或碟形)封头。这种封头形式在压力容器中很常用。其厚度常和与之相连的圆筒体厚度一样。这大大减少了焊接制造量。因此,由于折边区以外的区域所需的厚度小于封头的实际厚度,故多余的部分就可以用于这些区域内的接管补强。许多工厂能制成各种直径和厚度的这种封头,而且价格很有竞争力。

(4) 锥形和准锥形封头。这种封头用于圆锥体和塔器,可作为底部封口或用作不同直径圆筒体之间的过渡区域。因为圆锥与圆筒体连接区域的高不平衡力,所以该区域必须视作圆锥设计的一部分。由于这些高不平衡力,因此,ASME 规范第八篇第一分篇限制了受内压圆锥体锥顶角最大值为 30°。大于 30° 的必须对其进行不连续应力分析或采用准锥形封头,以避免连接区产生不平衡力。

3. 盲板法兰、平盖板和法兰

压力容器最常用的封头形式之一是不固定的平封头或平盖。它可以与壳体制成整体,也可以与壳体相焊,或用螺栓或一些快开装置与壳体相连。它可以是圆的、方的、矩形的或其他一些形状。那种用螺栓连接并采用密封垫片的圆平盖封头称作盲板法兰。

通常盲板法兰用螺栓与容器法兰相连,并在两法兰之间使用密封垫片。虽然平封头或盲板法兰可用圆形或非圆形的,但通常都有相同的厚度。

4. 开孔和接管

所有的压力容器必须开孔,以便于内部所有之物的进和出。一些内件很大或需要频繁更换内件的容器,则去除整个封头或壳体的一部分,以制成大开孔作为通道。而对于多数容器,内件可通过封头或壳体上的开孔进出,开口处有接管和管道与之相连。除了这些开孔,还需要有其他的开孔,例如便于人员进入容器的人孔。其他开孔也可能是需要的:通过手孔从容器外部检查容器,以及用于清洁和吹干容器的其他开孔。这些开孔并非总是在开孔处有接管与之相连。有些封口处是一个人孔盖或手孔盖,可以与容器直接相焊或用螺栓相连接。

5. 支座

多数直立容器用裙座支撑。裙座通过剪切作用转移容器的载荷而很经济。裙座同样通过地脚螺栓和基础环将载荷转移给底部。支腿式支撑的容器重量通常很轻,并且支腿便于(人员)进入容器底部。一种经济的设计是支腿直接与容器相连。通过剪切作用转移载荷。水平放置的容器通常采用鞍式支座。如果容器壁太薄以至于不能将载荷转移至鞍座,则必须采用加强圈,还必须考虑热膨胀问题。

Unit 20

Design of Pressure Vessels

1. Selection of Vessel

Although many factors contribute to the selection of pressure vessels, the two basic requirements that affect the selection are safety and economics. Many items are considered, such as materials' availability, corrosion resistance, materials' strength, types and magnitudes of loadings, location of installation including wind loading and earthquake loading, location of fabrication—shop or field, position of vessel installation, and availability of labor supply at the erection site.

With increasing use of special pressure vessels in the petrochemical and other industries, the availability of the proper materials is fast becoming a major problem. The most usual material for vessels is carbon steel. Many other specialized materials are also being used for corrosion resistance or the ability to contain a fluid without degradation of the material's properties. Substitution of materials is prevalent, and cladding and coatings are used extensively. The design engineer must be in communication with the process engineer in order that all materials used will contribute to the overall integrity of the vessel. For those vessels that require field assembly in contrast to those that can be built in the shop, proper quality assurance must be established for acceptable welding regardless of the adverse conditions under which the vessel is made. Previsions must be established for radiography, stress relieving, and other operations required in the field.

For those vessels that will operate in climates where low temperatures are encountered or contain fluids operating at low temperatures, special care must be taken to ensure impact resistance of the materials at low temperatures[1]. To obtain this property, the vessel may require a special high-alloy steel, nonferrous material, or some special heat treatment.

2. Which Pressure Vessel Code Is Used

The first consideration must be whether or not there is a pressure vessel law at the location of the installation. If there is, the applicable codes are stated in the law. If the jurisdiction has adopted the ASME Code, Section Ⅷ, the decision may be narrowed down to selecting whether Division 1 or Division 2 is used[2].

There are many opinions regarding the use of Division 1 versus Division 2, but the "bottom line" is economics. Division 1 uses approximate formulas, charts, and graphs in simple calculations. Division 2, on the other hand, uses a complex method of formulas, charts, and design-by-analysis which must be described in a stress report. Sometimes so many additional requirements are added to the minimum specifications of a Division 1 vessel that it might be more economical to supply a Division 2 vessel and take advantage of the higher allowable stresses[3].

3. Special Design Requirements

In addition to the standard information required on all units, such as design pressure, design temperature, geometry, and size, many other items of information are necessary and must be recorded. The corrosion and erosion amounts are to be given and a suitable material and method of protection are to be noted. The type of fluid that will be contained, such as lethal, must be noted because of the required specific design details. Supported position, vertical or horizontal, and support locations must be listed as well as any local loads from supported equipment and piping. Site location is given so that wind, snow, and earthquake requirements can be determined. Impact loads and cyclic requirements are also included.

4. Design Reports and Calculations

The ASME Code, Ⅷ-2, requires a formal design report with the assumptions in the User's Design Specification incorporated in the stress analysis calculations. These calculations are prepared and certified by a registered professional engineer experienced in pressure vessel design. As with the User's Design Specification, the Manufacturer's Design Report is mandatory and the certification reported on the Manufacturer's Data Report[4]. This is kept on file by the manufacturer for five year.

5. Materials' Specifications

All codes and standards have materials' specification and requirements describing which materials are permissible. Those materials that are permitted with a specific code are either listed or limited to the ones that have allowable stress values given. Depending upon the code or standard, permitted materials for a particular process vessel are limited. For instance, only materials with an SA or SB designation can be used in ASME Boiler and Pressure Vessel Code construction.

6. Factors of Safety

The factors of safety are directly related to the theories and modes of failure, the specific design criteria of each code, and the extent to which various levels of actual stresses are determined and evaluated.

Throughout the world, various factors of safety are applied to materials' data to establish allowable stresses for the design of boilers, pressure vessels, and piping.

For the temperature range to that temperature where creep or rupture sets the allowable stresses，the universal factor for setting allowable stresses is based on yield strength.

Words and Expressions

installation	[ˌɪnstəˈleɪʃn]	n.	安装,装置
earthquake	[ˈɜːθkweɪk]	n.	地震
erection	[ɪˈrekʃn]	n.	竖立,安装
petrochemical	[ˌpetrəʊˈkemɪkl]	adj.	石油化工的
degradation	[ˌdegrəˈdeɪʃn]	n.	降级,退化
substitution	[ˌsʌbstɪˈtjuːʃn]	n.	代替,取代
prevalent	[ˈprevələnt]	adj.	普遍的,流行的
integrity	[ɪnˈtegrəti]	n.	完整性
assembly	[əˈsembli]	n.	组合,装配
assurance	[əˈʃʊərəns]	n.	确信,保证,保险
adverse	[ˈædvɜːs]	adj.	不利的,有害的,敌对的
radiography	[reɪdiˈɒɡrəfi]	n.	射线照相,X 光探伤
encounter	[ɪnˈkaʊntə]	vt.	遭遇,碰到
jurisdiction	[ˌdʒʊərɪsˈdɪkʃn]	n.	权限,管辖权
erosion	[ɪˈrəʊʒn]	n.	腐蚀,冲刷
lethal	[ˈliːθl]	adj.	致命的,致死的
sustained	[səsˈteɪnd]	adj.	持续的,不衰减的
transient	[ˈtrænziənt]	adj.	瞬态的,过渡的
assumption	[əˈsʌmpʃn]	n.	假设,设想,前提
specification	[ˌspesɪfɪˈkeɪʃn]	n.	技术要求,详细说明;说明书,详细的计划书
certification	[ˌsɜːtɪfɪˈkeɪʃn]	n.	证明,确认,鉴定
criteria	[kraɪˈtɪəriə]	n.	判据;准则;标准;指标

NOTES

[1] For those vessels that will operate in climates where low temperatures are encountered or contain fluids operating at low temperatures，special care must be taken to ensure impact resistance of the materials at low temperatures.

对于那些在低温环境下运行或盛装低温液体的容器,必须特别注意保证材料在低温下的抗冲击能力。

［2］If the jurisdiction has adopted the ASME Code，Section Ⅷ，the decision may be narrowed down to selecting whether Division 1 or Division 2 is used.

如果管辖部门已决定采用 ASME 规范的第八篇，那么需要确定的只是选用第一分篇还是第二分篇。

［3］Sometimes so many additional requirements are added to the minimum specifications of a Division 1 vessel that it might be more economical to supply a Division 2 vessel and take advantage of the higher allowable stresses.

有时，由于对按第一分篇设计的容器在最低要求之外又增加了许多附加要求，因此，按第二分篇设计并选取较高许用应力可能更为经济。

［4］As with the User's Design Specification，the Manufacturer's Design Report is mandatory and the certification reported on the Manufacturer's Data Report.

如用户设计技术条件一样，制造厂的设计报告以及有关制造厂数据报告的证明书都是强制性的。（Manufacturer's Data Report 后面省略了 is mandatory）

第20单元　压力容器设计

1. 容器选择

虽然很多因素影响压力容器的选择，但是影响选择的两个基本必要因素是安全性和经济性。许多因素需要考虑，诸如材料的可获得性、材料的抗蚀性能、材料强度、载荷的形式和大小、安装地区情况（包括风载荷和地震载荷）、制造地点——工厂或是野外、容器安装的位置，以及建造地可选用的劳动力供应情况。

随着石油化工和其他工业中特殊压力容器使用的增加，获取合适的材料迅速成为主要问题。容器最通常采用的材料是碳钢，许多其他专门材料也被用来抗腐蚀或者贮存某流体而不退化材料性能。材料替代很普遍而且覆盖层和涂层材料也被广泛采用。设计工程师必须与工艺工程师保持联系，以便所采用的所有材料都将有助于容器的完整性。对于那些需要现场安装的容器，相对于在车间制造的容器来说，尽管容器是在不利的条件下制造的，但也必须制定适当的焊接质量保证。必须预先制定有关 X 射线探伤、应力消除和其他现场操作的要求。

对于那些在低温环境下运行或盛装低温液体的容器，必须特别注意保证材料在低温下的抗冲击能力，为了获得这种特性，容器可能需要特殊的高合金钢、非铁材料，或进行一些特殊的热处理。

2. 采用何种设计规范

首先必须考虑的是安装当地是否有压力容器法规，如果有，采用的规范必须是法规中规定的。如果管辖部门已决定采用 ASME 规范的第八篇，那么需要确定的只是选用第一分篇还是第二分篇。

关于选用第一分篇还是第二分篇有很多不同的观点，但是"底线"是经济性，第一分篇在简化计算中采用了近似的公式、图表和曲线图。另一方面,第二分篇采用了复杂的公式方法、图表和应力分析,而应力分析必须在应力报告中进行描述。有时,由于对按第一分篇设计的容器在最低要求之外又增加了许多附加要求,因此,按第二分篇设计并选取较高的许用应力可能更为经济。

3. 特殊设计要求

除了所有单元设备必备的,如设计压力、设计温度、几何形状和尺寸大小这些标准信息外,还需要很多其他的信息并且必须记录在案。需给出腐蚀量和磨损量,并且记录适合的材料和维护方法。因为特定设计资料的需要,所以必须记录所装流体的类型,例如致命性。必须列出垂直或水平的支座位置,同时列出来自所支撑的容器和管道的任何局部载荷,指明容器安置的地点,以便确定风载荷、雪载荷和地震载荷,还要列出冲击载荷和载荷循环要求。

4. 设计报告和计算书

ASME 规范第八篇第二分篇要求包含一篇正式的设计报告,内容须包含在应力分析计算中,对用户设计技术规范的假定条件。这些计算书由经注册的、富有经验的压力容器设计职业工程师准备和确认。如用户设计技术条件一样,制造厂的设计报告以及有关制造厂数据报告的证明书都是强制性的。这些文件由制造厂保留5年。

5. 材料技术特性

所有的规范和标准都列出了材料技术特性和允许采用的材料要求。那些特殊规范允许的材料或者被列出,或者仅限于给出许用应力值。根据规范和标准,对一个特定的容器来说,允许采用的材料是有限制性的。例如,在 ASME 锅炉及压力容器规范中指定制造只能采用带有 SA 或 SB 标记的材料。

6. 安全系数

安全系数直接与采用的理论和失效模式、每种规范特定的设计标准以及计算评估出的各种实际应力水平有关。

在全世界,锅炉、压力容器和管道设计时采用各种安全系数以确定材料的许用应力。确定许用应力的温度范围可达到蠕变或破裂时的温度,所以普遍采用基于屈服强度的安全系数来确定许用应力值。

READING MATERIAL

Types of Heat Exchangers

Heat exchangers are equipment primarily for transferring heat between hot and cold streams. They have separate passages for the two streams and operate continuously. The most versatile and widely used exchangers are the shell-and-tube types, but various plate and other types are valuable and economically competitive or superior in some applications. These types are largely proprietary and for the most part must be process designed by their manufacturers[1].

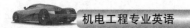

1. Plate-and-Frame Exchangers

Plate-and-frame exchangers are assemblies of pressed corrugated plates on a frame. Gaskets in grooves around the periphery contain the fluids and direct the flows into and out of the spaces between the plates. Close spacing and the presence of the corrugations result in high coefficients on both sides several times those of shell-and-tube equipment and fouling factors are low. The accessibility of the heat exchange surface for cleaning makes them particularly suitable for fouling services and where a high degree of sanitation is required, as in food and pharmaceutical processing. Operating pressures and temperatures are limited by the natures of the available gasketing materials, with usual maxima of 300 psig and $400°$F.

Since plate-and-frame exchangers are made by comparatively few concerns, most process design information about them is proprietary but may be made available to serious engineers[2]. Friction factors and heat transfer coefficients vary with the plate spacing and the kinds of corrugations.

2. Spiral Heat Exchangers

In spiral heat exchangers, the hot fluid enters at the center of the spiral element and flows to the periphery; flow of the cold liquid is countercurrent, entering at the periphery and leaving at the center. Heat transfer coefficients are high on both sides. These factors may lead to surface requirements 20% or so less than those of shell-and-tube exchangers. Spiral types generally may be superior with highly viscous fluids at moderate pressures.

3. Compact(Plate-Fin) Exchangers

Compact exchangers are used primarily for gas service. Typically they have surfaces of the order of $1\,200\ \text{m}^2/\text{m}^3$, corrugation height $3.8\sim11.8\ \text{mm}$, corrugation thickness $0.2\sim0.6\ \text{mm}$, and fin density $230\sim700\ \text{fins/m}$. The large extended surface permits about four times the heat transfer rate per unit volume that can be achieved with shell-and-tube construction. Units have been designed for pressures up to 80 atm or so. The close spacings militate against fouling service. Commercially, compact exchangers are used in cryogenic services, and also for heat recovery at high temperatures in connection with gas turbines. For mobile units, as in motor vehicles, compact exchangers have the great merits of compactness and light weight. Any kind of arrangement of cross and countercurrent flows is feasible, and three or more different streams can be accommodated in the same equipment[3]. Pressure drop, heat transfer relations, and other aspects of design are well documented.

4. Double-Pipe Exchangers

This kind of exchanger consists of a central pipe supported within a larger one by packing glands. The straight length is limited to a maximum of about 20 ft, otherwise the center pipe will sag and cause poor distribution in the annulus. It is customary to operate with the high pressure，high temperature，high density，and corrosive fluid in the inner pipe and the less demanding one in the annulus. The inner surface can be provided with scrapers as in dewaxing of oils or crystallization from solutions. External longitudinal fins in the annular space can be used to improve heat transfer with gases or viscous fluids. When greater heat transfer surfaces are needed，several double-pipes can be stacked in any combination of series or parallel.

5. Shell-and-Tube Exchangers

This type of exchanger has one shell-side pass and one tube-side pass. This construction，where one shell server for many tubes，is very economical. In an exchanger the shell-side and tube-side heat-transfer coefficients are of comparable importance，and both must be large if a satisfactory overall coefficient is to be attained. It is so important and so widely used in the process industries that it will be documented in detail in other literature.

Words and Expressions

passage	[ˈpæsɪdʒ]	n.	通道,过道
versatile	[ˈvɜːsətaɪl]	adj.	(指工具、机器等)多用途的,多功能的
proprietary	[prəˈpraɪətri]	adj.	专利的,私有的
corrugate	[ˈkɒrʊgeɪt]	vt.	(使某物)起皱褶,起皱纹,起波纹,成波纹状
coefficient	[ˌkəʊɪˈfɪʃənt]	n.	系数;(测定某种质量或变化过程的)率,程度
gasket	[ˈgæskɪt]	n.	密封垫片,垫圈,衬垫
foul	[faʊl]	vt.	使污秽;弄脏,堵塞
fouling factor			污垢系数
sanitation	[ˌsænɪˈteɪʃn]	n.	卫生,卫生系统或设备
pharmaceutical	[ˌfɑːməˈsjuːtɪkl]	adj.	制药的,配药的,药物的
countercurrent	[ˈkaʊntəˌkʌrənt]	adv.	相反地
fin	[fɪn]	n.	翅片;(汽车、飞机、炸弹上的)尾翅
militate	[ˈmɪlɪteɪt]	vt.	妨碍,起作用
cryogenic	[ˌkraɪəʊˈdʒenɪk]	adj.	冷冻的,低温的,低温学的
recovery	[rɪˈkʌvəri]	n.	恢复,回收,再生;恢复健康,复原
gland	[glænd]	n.	填料盖,密封套
sag	[sæg]	vi.	向下凹或中间下陷,下垂,下沉;松弛或不整

annulus	[ˈænjʊləs]	n.	环面,环状空间
dewax	[diːˈwæks]	vt.	使脱蜡
crystallization	[ˌkrɪstəlaɪˈzeɪʃn]	n.	结晶,结晶体
stack	[stæk]	n.	堆积,垛;大量,一大堆;大烟囱;(船上的)烟囱
inherent	[ɪnˈherənt]	adj.	固有的;内在的
accommodate	[əˈkɒmədeɪt]	vt.	使适应,调节;容纳;向……提供住处

NOTES

[1] These types are largely proprietary and for the most part must be process designed by their manufacturers.

这些类型的换热器基本上是带有专利性的,并且多数必须由它们的制造厂进行工艺设计。

[2] Since plate-and-frame exchangers are made by comparatively few concerns, most process design information about them is proprietary but may be made available to serious engineers.

因为只有较少企业生产板框式换热器,所以大多数关于板框式换热器的工艺设计资料是带有专利性的,但可以提供给重要的工程师。(A is available to B,B 可以得到 A。句中的 concerns 可译为"企业""财团")

[3] Any kind of arrangement of cross and countercurrent flows is feasible, and three or more different streams can be accommodated in the same equipment.

错流和逆流的任何排列形式都是可行的,并且在同一设备中可以安排3种或多种流束。

阅读材料

换热器类型

换热器是一种主要用于冷、热流体间交换热量的设备。两种流体在换热器中有独立通道并可连续工作。用途最广、应用最广泛的换热器是管壳式换热器。但是在某些场合,各种板式和其他类型的换热器具备经济竞争力或者性能更优越。这些类型的换热器基本上是带有专利性的,并且多数必须由它们的制造厂进行工艺设计。

1. 板框式换热器

板框式换热器是在框内安装压制成波纹状的薄片而成。板片四周的槽内安装垫片,以盛装流体(不泄漏),并引导流体在板与板之间的间隙内流进、流出。封闭的空间和波纹片的使用使得换热器在板片两侧具有较高的传热效率,是管壳式设备的数倍,并且污垢系数低。这种换热器,由于换热表面易于清洗,尤其适合于介质不清洁和卫生要求高的场合,如食品和制药工业。使用压力和使用温度受密封垫片材料性能的限制,不宜过高,一般最大不超过300磅/英寸和400°F。

因为只有较少企业生产板框式换热器,所以大多数关于板框式换热器的工艺设计资料是带有专利性的,但可以提供给重要的工程师。摩擦系数和换热效率随着板间间隙和板面波纹形状不同而不同。

2. 螺旋式换热器

螺旋式换热器中,热流体由螺旋件的中心部位进入再流向外圈,而冷流体逆向而行,由外圈进入换热器再由中心部位流出。两侧的传热效率高。这些因素导致其所需换热面积只是管壳式换热器的20%或更少。在适当的压力下,对于高黏度流体的换热,采用螺旋式换热器尤为适合。

3. 紧凑式(板翅式)换热器

紧凑式换热器主要用于处理气体介质,其特点是:单位体积内传热面积大,可达到 $1\,200\ m^2/m^3$,波纹片高度 $3.8\sim11.8\ mm$,波纹片厚度 $0.2\sim0.6\ mm$,波纹片密度 $230\sim 700$ 翅片/米。大大增加的传热面积使得其单位体积的传热效率是管壳式结构的 4 倍,设备设计压力可达到 80 个大气压左右。封闭的空间使其不宜用于较脏的场合。商业上,紧凑式(板翅式)换热器用于低温操作,也可与汽轮机相连接在高温下回收热量。对于汽车、摩托车也一样,紧凑式换热器由于结构紧凑、重量轻而具有明显优势。错流和逆流的任何排列形式都是可行的,并且在同一设备中可以安排 3 种或多种流束。关于压力降、热量传递和设计的其他方面都已有很好的论述。

4. 双(套)管式换热器

这种换热器是在大管内套同心小管,再包扎密封而成。直管部分长度不超过大约 20 英尺,否则中心管会下沉并造成环形通道内流体分布不均。操作时,高压、高温、高密度和腐蚀性的流体常走内管,对材料要求低的走外围环形通道。在石油脱蜡和溶体结晶工艺中,换热器内表面容易刮伤。对于气体和黏性流体,外围环状通道内采用轴向翅片可提高换热效率。如果需要更大的换热面积,可将几组套管串联或并联组合使用。

5. 管壳式换热器

这种类型的换热器具有一个壳程和一个管程。这种结构中,一个壳体当作很多管子使用,所以非常经济。在换热器中,壳壁与管壁的传热系数较为重要,要达到令人满意的总传热系数,两者都必须较大。这种换热器在过程工业中的应用相当重要和广泛,因而将在别的文献中做详细论述。

Unit 21

Basic Packed Tower Design

A common apparatus used in gas absorption and certain other operations is the packed tower. The device consists of a cylindrical column or tower, equipped with a gas inlet and distributing space at the bottom; a liquid inlet and distributor at the top; gas and liquid outlets at the top and bottom, respectively; and a supported mass of inert solid shapes, called tower packing.

In comparison with tray towers, packed towers are suited to small diameters(24 in or less), whenever low pressure is desirable, whenever low holdup is necessary, and whenever plastic or ceramic construction is required. Applications unfavorable to packings are large diameter towers, especially those with low liquid and high vapor rates, because of problems with liquid distribution, and whenever high turndown is required[1]. In large towers, random packing may cost more than twice as much as sieve or valve trays.

Depth of packing without intermediate supports is limited by its deformability: metal construction is limited to depths of $20 \sim 25$ ft, and plastic to $10 \sim 15$ ft. Intermediate supports and liquid redistributors are supplied for deeper beds and at sidestream withdrawal or feed points. Liquid redistributors usually are needed every $2.5 \sim 3$ tower diameters for Raschig rings and every $5 \sim 10$ diameters for Pall rings, but at least every 20 ft.

The various kinds of internals of packed towers are described briefly below:

A combination packing support and redistributors can support packings for deeper beds and also serve as a sump for withdrawal of the liquid from the tower.

A trough-type distributor is suitable for liquid rates in excess of 2 gpm/sqft in towers 2 feet and more in diameter. They can be made in ceramics or plastics.

A perforated pipe distributor is available in a variety of shapes, and is the most efficient type over a wide range of liquid rates; in large towers and where distribution is especially critical, they are fitted with nozzles instead of perforations.

A redistribution device, the rosette, provides adequate redistribution in small diameter towers; it diverts the liquid away from the wall towards which it tends to go.

A hold-down plate can keep low density packings in place and to prevent fragile packings such as those made of carbon, for instance, from disintegrating because of mechanical disturbances at the top of the bed.

The broad classes of packings for vapor-liquid contacting are either random or structured[2]. The former are small, hollow structures with large surface per unit volume that are loaded at random into the vessel. Structured packings may be stacked in layers of large rings or grids or as spiral windings.

There are several kinds of packings. The first of the widely used random packings were Raschig rings which are hollow cylinders of ceramics, plastics, or metal. They were an economical replacement for the crushed rock often used then. Because of their simplicity and their early introduction, Raschig rings have been investigated thoroughly and many data of their performance have been obtained which are still useful.

Structured packings are employed particularly in vacuum service where pressure drops must be kept low. Because of their open structure and large specific surface, their mass transfer efficiency is high when proper distribution of liquid over the cross section can be maintained.

Most tower packings are made of cheap, inert, fairly light materials such as clay, porcelain, or various plastics. Thin-walled metal rings of steel or aluminum are sometimes used. High void spaces and large passages for the fluids are achieved by making the packing units irregular or hollow, so that they interlock into open structures with a porosity of 60~95 percent.

Words and Expressions

column	[ˈkɒləm]	n.	柱,圆柱;塔,塔体
tray	[treɪ]	n.	盘
tray tower			板式塔
holdup	[həʊldʌp]	n.	滞留量,塔储量
ceramic	[sɪˈræmɪk]	adj.	陶瓷的,陶器的,与陶器有关的
packing	[ˈpækɪŋ]	n.	填料,填充物
turndown	[ˈtɜːndaʊn]	n.	调节,翻折物
sieve	[sɪv]	n.	筛,筛网
sieve tower			筛板塔
valve tray			浮阀塔板,浮阀盘
random	[ˈrændəm]	adj.	随意的,偶然的;任意的;无计划的
deformability	[dɪˌfɔːməˈbɪlɪti]	n.	可变形能力,形变度,可塑性

redistributor	[ˌriːdisˈtrɪbjʊtə]	n.	再分配器
sidestream	[ˈsaɪdstriːm]	n.	侧流,边流
withdrawal	[wɪðˈdrɔːəl]	n.	回收,取回,去除
Raschig rings			拉西环
Pall rings			鲍尔环
sump	[sʌmp]	n.	集液盘,污水坑,水坑,池,机油箱
trough	[trɒ(ː)f]	n.	槽,沟,引水道
perforate	[ˈpɜːfəreɪt]	vt.	穿孔于,在……上打眼
rosette	[rəʊˈzet]	n.	玫瑰花形物
fragile	[ˈfrædʒaɪl]	adj.	易碎的,脆的;虚弱的,脆弱的,经不起折腾的
disintegrate	[disˈɪntɪɡreɪt]	vt. & vi.	(使)破裂;分裂,粉碎,(使)崩溃
winding	[ˈwaɪndɪŋ]	n.	绕组,线圈
porcelain	[ˈpɔːsəlɪn]	n.	瓷,瓷器
porosity	[pɔːˈrɒsəti]	n.	多孔性,有孔性,孔隙率
interlock	[ˌɪntəˈlɒk]	vt.	联结;互锁

NOTES

[1] Applications unfavorable to packings are large diameter towers, especially those with low liquid and high vapor rates, because of problems with liquid distribution, and whenever high turndown is required.

尤其当气体比例高、液体比例低时,由于液体分布的问题,以及需要较大的调节能力,所以填料塔不适合用大直径。

[2] The broad classes of packings for vapor-liquid contacting are either random or structured.

气液在其中接触的填料分为散装填料和规整填料两大类。(random packing 可译为"散装填料""乱堆填料",structured packing 可译为"规整填料""整砌填料")

第21单元　填料塔设计基础

填料塔是一种用于气体吸收及一些其他操作的普通设备。这种装置由一个圆柱形竖筒塔架组成,在底部装有一个进气口和分布空间;顶部有一个液体入口和分布器;气体和液体出口分别在塔的顶部和底部,被支承的惰性固体物质称为塔填料。

与板式塔相比,每当需要压力较低、滞留量较少和要求采用塑料或陶瓷结构时,填料塔更适合于小直径(24英寸或更小)的场合。尤其当气体比例高、液体比例低时,由于液体分布的问题,以及需要较大的调节能力,所以填料塔不适合用大直径。在大直径塔内,散装填料的成本可能是筛板和浮阀盘的2倍以上。

无中间支承的填料深度受填料变形能力的限制;金属结构的填料深度仅限于20～25英尺,塑料的不超过10～15英尺。在壁流回收处或进料口处可设置中间支承和液体再分布器,来满足更深填料的要求。对拉西环填料而言,通常每隔2.5～3倍的塔径间距,设置一个液体再分布器;而对鲍尔环而言,每隔5～10倍的塔径,但至少每隔20英尺设置一个。

对填料塔的各种内件做简要描述:

填料支承与液体再分布器的组合装置不仅可增加填料层深度,而且可用作集液盘,用以回收从上层塔壁流下的液体。

槽式分布器适用于直径2英尺及以上的塔,液体流量超过2加仑每分钟/平方英尺的场合,其材料可用陶瓷或塑料。

孔管式分布器可制成多种形状,并且是在处理宽泛的液体流量范围内最有效的类型。在大直径塔和严格要求分布效果时,它们安装了喷嘴,而不是在上面穿孔。

液体再分布器呈玫瑰花形,可在小直径塔内充分重新分布液体,能将沿塔壁流下的液体导离塔壁、往设计好的方向流动。

填料支承盘可支撑较稀疏的填料,且可防止如碳钢等制的易碎填料由于床层顶部的机械作用而被粉碎。

气液在其中接触的填料分为散装填料和规整填料两大类。前者个体体积较小、具有中空结构且单位体积的表面积较大,并以乱堆方式安装在容器内。规整填料以大圆环或网状或似螺旋线圈状整砌而成。

填料的种类很多。应用最广泛的散装填料是拉西环,其结构为中空的圆柱体,材料可为陶瓷、塑料或金属。它们用来代替以前常用的碎石子填料,非常经济。拉西环因其结构简单且较早使用,所以对它的研究较为透彻,并且关于它的性能已取得较多数据且仍在使用这些数据。

规整填料尤其适合于需要维持较低的压力降的真空设备,由于多孔的结构和大的表面积,当液体在(填料)横截面的分布维持恰当时,可获得较高的传质效率。

大多数填料是由廉价、稳定且质轻的材料制成,例如黏土、陶瓷或各种塑料。有时也用钢制或铝制的薄壁金属环。将填料单元制成不规则或中空的形状,可使流动获得高空隙空间和大的通道,因此,填料一般都连接成孔隙率可达60%～95%的多孔结构。

READING MATERIAL

Tray Efficiency

When designing a new distillation column or revamping an existing one, it is important to understand the factors that affect overall tray efficiency.

Despite decades of work on tray efficiency, it is still unclear which design variables affect tray efficiencies of commercial columns and to what extent. Many of the efficiency studies have been contradictory, and others relied on data derived from laboratory-scale towers that do not extend to commercial towers. Based on much experience gathered in the industry over the years, and a large amount of commercial-scale data recently released by Fractionation Research Inc.(FRI), it is now possible to get a clearer picture of the factors that affect overall tray efficiency[1].

Previous work showed that errors in vapor-liquid equilibrium(VLE) data and reflux ratio measurements can have a major effect on tray efficiency calculations[2]. Past work also showed that overall tray efficiency increases with lower viscosities and relative volatilities. This article presents an analysis of FRI data and quantitatively defines the effects of tower geometry(flow path length, fractional hole area, hole diameter, and weir height) on overall tray efficiency in commercial-scale fractionators.

The pitfalls of obtaining tray efficiency from test data:

(1) Overall column efficiency. The overall column(or overall tray) efficiency, EOC, is the ratio of the number of theoretical stages required for the separation[not counting the reboiler(s) and condenser(s)] to the number of trays in the tower. Since tray efficiencies may vary from one tower section to another, the overall column efficiency concept is often applied separately to the stripping section and the rectifying section. It is easy to apply, and generally provides a good characterization of tray efficiencies in commercial towers. This concept is, therefore, preferred by most industrial practitioners. Alternative definitions of tray efficiency are more often employed to relate tray efficiency to masstransfer fundamentals.

Overall column efficiency is normally obtained from test data by matching a stage-to-stage calculation, usually by means of a commercial tower simulation, to test data. Based on the measured mass and component balances, boilup, reflux, temperatures, pressures and compositions, the number of stages in the calculation is varied until the closest match to data is obtained.

(2) Reliability of field and test data. Deriving reliable data from industrial operating towers is difficult. Tabulations of measured efficiency data obtained from operating columns are available, but their limitations and reliability are not always well-defined.

An excellent source of commercial-scale test data is measurements by FRI. For many years, only limited FRI data have been published in the open literature, but recently much more data have been released. The FRI tests were conducted in a research facility that permitted close monitoring, using standard test systems whose physical properties are well-known. This arrangement overcomes many of the limitations encountered in the testing of operating plant towers. Most of the data discussed here was obtained by FRI.

(3) Accuracy of VLE correlations. Errors in the VLE relationships are the result of inaccuracies in VLE test data，and are known only within limits based on the source of the VLE information. This gives rise to uncertainty in both the VLE data and the resulting stage-to-stage calculations.

Most efficiency test data reported in the literature were obtained at total reflux. This is true of most of FRI's published data as well as most of the other data derived from test columns. The total-reflux test mode is often preferred，because measurements under total reflux do not suffer from compounding of inaccuracies due to finite reflux ratios，nor do they suffer from inaccuracies in mass and energy balances. Reflux ratio has been reported to have a small effect on tray efficiency.

Words and Expressions

distillation	[ˌdɪstɪˈleɪʃn]	n.	蒸馏(过程);蒸馏物
revamp	[ˈriːvæmp]	vt.	修补
commercial-scale	[kəˈmɜːʃəl skeɪl]	adj.	工业规模(的),大规模(的)
reflux	[ˈriːflʌks]	n.	逆流,回流,退潮
total reflux			全回流
viscosity	[vɪsˈkɒsəti]	n.	黏稠;黏性
volatility	[ˌvɒləˈtɪlɪti]	n.	挥发性
fractional	[ˈfrækʃənl]	adj.	微不足道的,极小的,极少的
fractionator	[ˌfrækʃəneɪtə]	n.	分馏器
fractionation	[ˈfrækʃəˈneɪʃn]	n.	分别,分馏法
pitfalls	[ˈpɪtfɔːlz]	n.	意想不到的困难,易犯的错误
reboiler	[rɪˈbɔɪlə]	n.	再沸器,重沸器,再煮器[锅]
condenser	[kənˈdensə]	n.	冷凝器
strip	[strɪp]	vi. & vt.	剥去,脱去,除去;剥夺,夺走
stripping section			提馏[反萃,剥淡,贫化]段
rectify	[ˈrektɪfaɪ]	vt.	精馏
rectifying section			精馏段
tabulation	[ˌtæbjʊˈleʃn]	n.	作表,表格
monitoring	[ˈmɒnɪtərɪŋ]	n.	监视,控制,监测,追踪
correlation	[ˌkɒrəˈleɪʃn]	n.	相互的关系
Fractionation Research Inc.			分馏研究公司

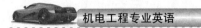

vapor-liquid equilibrium(VLE)　　　　　　气液平衡
commercial column　　　　　　　　　　　商业用塔,工业用塔

NOTES

[1] Based on much experience gathered in the industry over the years，and a large amount of commercial-scale data recently released by Fractionation Research Inc.(FRI)，it is now possible to get a clearer picture of the factors that affect overall tray efficiency.

基于过去数年工业上的许多经验以及分馏研究公司(FRI)最近发布的大量的工业规模塔的资料,现在有可能更清楚地了解影响全塔效率的因素。

[2] Previous work showed that errors in vapor-liquid equilibrium(VLE) data and reflux ratio measurements can have a major effect on tray efficiency calculations.

以前的工作显示出气液平衡(VLE)数据中的错误和测量回流比例中的错误主要影响了全塔效率的计算。(该句为主从复合句,主句是 Previous work showed… ,从属连词 that 引导的名词性从句 that errors in… and… calculations 充当主句的宾语,and 后面省略了 errors in)

阅读材料

塔板效率

当设计一个新蒸馏塔或修补现存塔时,了解影响全塔效率的因素是重要的。

尽管研究塔板效率已有数十年,但是影响工业用塔的塔板效率的设计变数及其影响程度仍然是含糊的。许多关于效率的研究是相互矛盾的,并且其他依赖实验室规模的塔数据的研究工作又难以推广到工业用塔。基于过去数年工业上的许多经验以及分馏研究公司(FRI)最近发布的大量的工业规模塔的资料,现在有可能更清楚地了解影响全塔效率的因素。

以前的工作显示出气液平衡(VLE)数据中的错误和测量回流比例中的错误主要影响了全塔效率的计算。过去的工作也表明随着黏性的降低和相对挥发度的降低,全塔效率提高了。本文介绍对 FRI 数据的分析,并且对工业规模用分馏塔全塔效率的几何影响因素(塔的流道长度、小孔面积、孔直径和堰高度)作一个定量规定。

从试验数据得到塔板效率的困难:

(1) 全塔效率。全塔效率(EOC)是指分离所需的理论级数(不含再沸器和冷凝器)与塔内的塔板数的比值。因为塔板效率会随各段塔节不同而不同,所以全塔效率的概念常分别用于提馏段和精馏段。它易于运用,且常对工业用塔的塔板效率提供很好的特性说明。因此,这个概念首先由大多数工业从业者提出。塔板效率的另一种定义更是常常将塔板效率与传质基本原则联系起来。

全塔效率通常由试验数据通过与试验数据匹配的逐级计算而得到,常采用模拟工业用塔测试数据的方法。在已测得的质量和各平衡组分(沸腾量、回流量、温度、压力和组成成分)的基础上,计算中的级数是变化的,直到与数据最为匹配为止。

(2) 实地和试验数据的可靠性。从工业操作塔获得可靠数据是困难的。从操作塔测算出的效率数据表格可在文献中获得,但是它们的局限性和可靠性一直没有明确的定义。

FRI 对优良的、源于工业规模的试验数据进行了测算。多年来,公开文献上只发表了 FRI 的数据,但是最近也发布了更多的数据。FRI 试验在一个研究设备中进行,该设备提供了紧密跟踪监测,采用的是标准试验系统,其物理性能众所周知。这种安排克服了工厂操作塔试验时遇到的许多局限性。在此讨论的多数数据由 FRI 获得。

(3) 气液平衡(VLE)关系的准确性。气液平衡(VLE)关系中的错误在于气液平衡(VLE)试验数据结果的不准确性,并且仅局限在基于 VLE 信息源的数据,这一点为人所熟知。这就导致了 VLE 数据和逐级计算结果的不确定性。

文献中报道的大多数有关效率的试验数据是在全回流时得到的。FRI 发表的大多数数据确实是这样的,同时由试验塔得到的大多数其他数据也是如此。全回流试验模型常常是较好的模型,因为在全回流下的测量不会遇到因有限回流率而引起的不正确的混合,也不会遇到不正确的质量和能量平衡。据报道,回流率对塔板效率的影响很小。

Unit 22

Types of Reactors

扫码可见本项目
☞ 参考资料

This section is devoted to the general characteristics of the main kinds of reactors.

The most obvious distinctions are between nonflow(batch) and continuous operating modes and between the kinds of phases that are being contacted. Batch processing is used primarily when the reaction time is long or the required daily production is small. The same batch equipment often is used to make a variety of products at different times. Otherwise, it is impossible to generalize as to the economical transition point from batch to continuous operation[1]. One or more batch reactors together with appropriate surge tanks may be used to simulate continuous operation on a daily or longer basis.

When heterogeneous mixtures are involved, the conversion rate often is limited by the rate of interphase mass transfer, so that a large interfacial surface is desirable. Thus, solid reactants or catalysts are finely divided, and fluid contacting is forced with mechanical agitation or in packed or tray towers or in centrifugal pumps. The rapid transfer of reactants past heat transfer surfaces by agitation or pumping also enhances heat transfer and reduces harmful temperature gradients.

1. Stirred Tanks

Stirred tanks are the most common type of batch reactor. Stirring is used to mix the ingredients initially, to maintain homogeneity during reaction, and to enhance heat transfer at a jacket wall or internal surfaces. Many applications of stirred tank reactors are to continuous processing, either with single tanks or multiple arrangements. Knowledge of the extent to which a stirred tank does approach complete mixing is essential to being able to predict its performance as a reactor[2].

2. Tubular Flow Reactors

The ideal behavior of tubular flow reactors(TFR) is plug flow, in which all nonreacting molecules have equal residence times. Any backmixing that occurs is incidental, the result of natural turbulence or that induced by obstructions to flow by catalyst granules or tower packing or necessary internals of the vessels[3]. The action of such obstructions can be two-edged, however, in that some local backmixing may occur，but on the whole a good approach to plug flow is developed because large scale turbulence is inhibited[4]. As a result of chemical reaction，gradients of concentration and temperature are developed in the axial direction of TFRs.

3. Gas-Liquid Reactors

Except with highly volatile liquids，reactions between gases and liquids occur in the liquid phase，following a transfer of gaseous participants through gas and liquid films. The rate of mass transfer always is a major or limiting factor in the overall transformation process. Naturally the equipment for such reactions is similar to that for the absorption of chemically inert gases，namely towers and stirred tanks.

4. Fixed Bed Reactors

The fixed beds of concern here are made up of catalyst particles in the range of 2~5 mm dia. The catalyst in a reactor may be loaded in several ways，as：

(1) a single large bed；

(2) several horizontal beds；

(3) several packed tubes in a single shell；

(4) a single bed with imbedded tubes；

(5) beds in separate shells.

5. Moving Bed Reactors

In such vessels，granular or lumpy material moves vertically downward as a mass. The solid may be a reactant or a catalyst or a heat carrier.

6. Fluidized Bed Reactors

This term is restricted here to equipment in which finely divided solids in suspension interact with gases. Solids fluidized by liquids are called slurries. Three phase fluidized mixtures occur in some coal liquifaction and petroleum treating processes. In dense phase gas-solid fluidization，a fairly definite bed level is maintained；in dilute phase systems，the solid is entrained continuously through the reaction zone and is separated out in a subsequent zone[5].

Words and Expressions

batch	[bætʃ]	adj.	分批的,间歇(式)的
surge	[sɜːdʒ]	n.	激增;大量
surge tank			平衡箱,调压槽
heterogeneous	[ˌhetərəˈdʒiːniəs]	adj.	非均质的,多相的;混杂的,多种多样的
reactant	[rɪˈæktənt]	n.	反应物
catalyst	[ˈkætəlɪst]	n.	[化]催化剂,触媒

stir	[stɜː]	vt. & vi.	搅拌
stirred tank			搅拌槽
homogeneity	[ˌhɒmədʒəˈniːəti]	n.	同种,同质,同质性,均匀性
jacket	[ˈdʒækɪt]	n.	夹套,套筒
tubular	[ˈtjuːbjʊlə]	adj.	管的;管(状)的,筒式的
backmix	[ˈbækmɪks]	n.	返混
obstruction	[əbˈstrʌkʃn]	n.	障碍物,阻碍物,阻碍,阻塞
granule	[ˈɡrænjuːl]	n.	小颗粒,小硬粒微粒
plug	[plʌɡ]	n.	插头,插座;塞子,栓
plug flow			活塞流
fluidize	[ˈfluːɪdaɪz]	vt.	使液化,流(态)化
slurry	[ˈslɜːri]	n.	泥浆,浆;悬浮体,稀浆
liquifaction	[ˌlɪkwɪˈfækʃn]	n.	液化,稀释
dilute	[daɪˈljuːt]	adj.	稀释的,冲淡的
entrain	[ɪnˈtreɪn]	vt.	夹带,带走

NOTES

[1] Otherwise, it is impossible to generalize as to the economical transition point from batch to continuous operation.

否则,间歇操作不可能比连续操作来得经济。

[2] Knowledge of the extent to which a stirred tank does approach complete mixing is essential to being able to predict its performance as a reactor.

了解搅拌槽所达到的完全混合程度,对预测它作为反应器的性能来说是极为必要的。

[3] Any backmixing that occurs is incidental, the result of natural turbulence or that induced by obstructions to flow by catalyst granules or tower packing or necessary internals of the vessels.

出现的任何返混都是伴随性的,它是自然湍流或由触媒颗粒、填料或容器中的必要内件对流动的阻碍所引起的湍流的结果。(the result... 前面省略了 and it is)

[4] The action of such obstructions can be two-edged, however, in that some local backmixing may occur, but on the whole a good approach to plug flow is developed because large scale turbulence is inhibited.

然而,这种阻碍有着两种相反的作用,它可能引起某些局部返混,但由于抑制了大规模的湍流,所以使得流动在整体上更接近于活塞流。

[5] In dense phase gas-solid fluidization, a fairly definite bed level is maintained; in dilute phase systems the solid is entrained continuously through the

reaction zone and is separated out in a subsequent zone.

在密相气—固流化中,维持着非常确定的床层高度;而在稀相系统中,固体被连续带入并通过反应区域,而在随后的区域内被分离。

第22单元　反应器类型

本部分专门介绍反应器的主要类型及其特点。

反应器最明显的区别是间歇操作还是连续式操作以及介质接触反应的各个阶段。间歇式操作主要用于反应时间长或日产量要求小的场合。同一台间歇设备经常在不同时段生产(不同的)产品,否则,间歇操作不可能比连续操作来得经济。一个或多个间歇反应器通过合适的调节槽连接起来,可用于单日或更长时间的近似连续操作。

不同种类的混合物混在一起时,其转化速率受内在传质速率的限制,所以就需要大的界面面积。因此,反应物或催化剂又做进一步的细分,并且流体在机械搅拌的作用下或者在填料塔,或板式塔内,或者在离心泵内接触。反应物在搅拌或抽吸作用下通过换热表面迅速地转化也加强了换热和降低了有害的温度梯度。

1. 搅拌槽(釜)

搅拌槽是最常用的间歇式反应器。反应最初时,搅拌器用来混合各成分,在反应过程中维持均匀性,并且加强夹套或内表面的换热。搅拌槽在连续工艺中也有许多应用,可以采用单个槽或多个槽组合排列使用。了解搅拌槽所达到的完全混合程度,对预测它作为反应器的性能来说是极为必要的。

2. 管式反应器

理想的管式反应器(TFR)是活塞流型,即所有不反应分子有相同的停留时间。出现的任何返混都是伴随性的,它是自然湍流或由触媒颗粒、填料或容器中的必要内件对流动的阻碍所引起的湍流的结果。然而,这种阻碍有着两种相反的作用,它可能引起某些局部返混,但由于抑制了大规模的湍流,所以使得流动在整体上更接近于活塞流。作为化学反应的结果,TFR可获得沿轴线方向的浓度梯度和温度梯度。

3. 气液反应器

除非是极易挥发的液体,否则当气体成分穿过气液薄膜层时,气液之间的反应发生在液体阶段。传质速率总是整个转化过程中的主要因素或限制因素。这种设备的实质类似于化学惰性气体吸收器,即塔器和搅拌槽。

4. 固定床反应器

这种情况下的固定床是由直径为2~5 mm范围的催化剂颗粒组成的。反应器中的催化剂安装有如下几种形式:

(1) 单个大型床层

(2) 几个水平排列的床层

(3) 单壳体中有几组管束

(4) 嵌入式管子组成的单床层

（5）独立分开壳体中的床层

5. 移动床反应器

这种容器中，小颗粒或小块物料作为一整块垂直下降，固体可以是反应物或催化剂或载热体。

6. 流化床反应器

此处的流化床严格地说是指悬浮的固体颗粒与气体相互作用的设备。被液体流化的固体称为悬浮体。在一些煤稀释和石油处理工艺中，产生三相流化混合物。在密相气—固流化中，维持着非常确定的床层高度；而在稀相系统中，固体被连续带入并通过反应区域，而在随后的区域内被分离。

READING MATERIAL

The Reciprocating Compressor

The basic reciprocating compression element is a single cylinder compressing on only one side of the piston(single-acting). A unit compressing on both sides of the piston(double-acting) consists of two basic single-acting elements operating in parallel in one casting.

The reciprocating compressor uses automatic spring loaded valves that open only when the proper differential pressure exists across the valve[1]. Inlet valves open when the pressure in the cylinder is slightly below the intake pressure. Discharge valves open when the pressure in the cylinder is slightly above the discharge pressure.

Figure 22-1(a) shows the basic element with the cylinder full of atmospheric air. On the theoretical pv diagram (indicator card), point 1 is the start of compression. Both valves are closed.

Figure 22-1(b) shows the compression stroke, the piston having moved to the left, reducing the original volume of air with an accompanying rise in pressure. Valves remain closed. The pv diagram shows compression from point 1 to 2, and that the pressure inside the cylinder has reached that in the receiver.

Figure 22-1(c) shows the piston completing the delivery stroke. The discharge valves opened just beyond point 2. Compressed air is flowing out through the discharge valves to the receiver.

After the piston reaches point 3, the discharge valves will close, leaving the clearance space filled with air at discharge pressure. During the expansion stroke, Figure 22-1(d), both the inlet and discharge valves remain closed and air trapped in the clearance space increases in volume causing a reduction in pressure[2]. This continues, as the piston moves to the right, until the cylinder pressure drops below the inlet pressure at point 4. The inlet valves now will open and air will flow into the cylinder until the end of the reverse stroke at point 1. This is the intake or suction stroke, illustrated by Figure 22-1(e). At point 1 on the pv diagram, the inlet valves will close and the cycle will repeat on the next revolution of the crank.

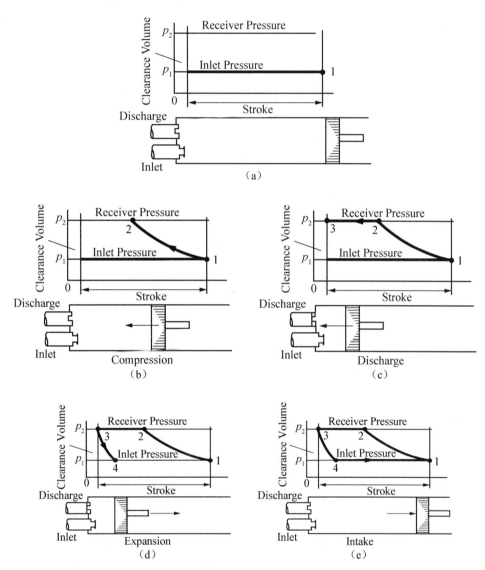

Figure 22 - 1　The various steps in a reciprocating compressor cycle

In a simple two-stage reciprocating compressor, the cylinders are proportioned according to the total compression ratio, the second stage being smaller because the gas, having already been partially compressed and cooled, occupies less volume than at the first stage inlet.

Looking at the pv diagram (Figure 22 – 2), the conditions before starting compression are points 1 and 5 for the first and second stages, respectively; after compression，points 2 and 6，and，after delivery，3 and 7. Expansion or air trapped in the clearance spaces as the pistons reverse brings points 4 and 8，and on the intake stroke the cylinders are again filled at points 1 and 5 and the cycle is set for repetition.

Multiple staging of any positive displacement compressor follows the above pattern.

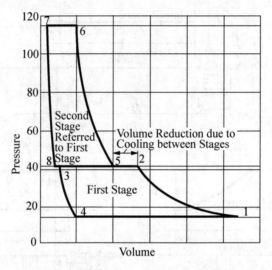

Figure 22‐2   Combined theoretical indicator card for a two-stage two-element 100 psiG positive-displacement compressor

Words and Expressions

reciprocating	[rɪˈsɪprəkeɪtɪŋ]	adj.	往复的,交替的,互换(的);摆动的
compressor	[kəmˈpresə]	n.	压气机,压缩机
piston	[ˈpɪstən]	n.	[机]活塞,柱塞
loaded	[ˈləʊdɪd]	adj.	带负载的,加重的
atmospheric	[ˌætməsˈferɪk]	adj.	大气的,大气层的;常压的
theoretical	[ˌθɪəˈretɪkl]	adj.	理论(上)的,假设的
accompany	[əˈkʌmpəni]	vt.	陪伴,陪同;伴随……同时发生
expansion	[ɪkˈspænʃn]	n.	膨胀,延伸率
inlet	[ˈɪnlet]	n.	进(气,水)口,入口
revolution	[ˌrevəˈluːʃn]	n.	循环,周期
delivery	[dɪˈlɪvəri]	n.	输送(气,水),排(气,水),排气(水)量
positive-displacement compressors			容积式压缩机
sliding-vane compressor			滑片式压缩机

liquid-piston compressor	液环式压缩机
two-impeller straight-lobe compressor	罗茨式压缩机
helical- or spiral-lobe compressor	螺杆式压缩机
dynamic compressor	速度式压缩机
centrifugal compressor	离心式压缩机
axial-flow compressor	轴流式压缩机
mixed-flow compressor	混流式压缩机

NOTES

[1] The reciprocating compressor uses automatic spring loaded valves that open only when the proper differential pressure exists across the valve.

往复式压缩机用一些仅在内外有一定压差存在时才开启的自动弹簧加载阀。(该句为主从复合句。that open only... 中的 that 引导定语从句修饰主句中的 valves, when the proper differential pressure exists across the valve 为状语从句,在定语从句中表示时间状语)

[2] During the expansion stroke, Figure 22-1(d), both the inlet and discharge valves remain closed and air trapped in the clearance space increases in volume causing a reduction in pressure.

在膨胀冲程中,如图 22-1(d)所示,进排气阀保持关闭并且封闭在余隙空间里的空气体积增加导致压力降低。(该句为并列复合句。During the expansion stroke 是介词短语作状语。前一句主语 valves 为复数,谓语是 remain;后一句主语是 air,过去分词短语 trapped in the clearance space 为主语的后置定语,谓语是 increases,介词短语 in volume 在原文中为状语,分词短语 causing a reduction in pressure 为结果状语)

阅读材料

往复式压缩机

基本往复式压缩机单元是一个仅在活塞一端压缩(单作用)的单个气缸。活塞两端压缩(双作用)单元由两个在一个壳体中平行操作的基本单作用单元组成。

往复式压缩机用一些仅在内外有一定压差存在时才开启的自动弹簧加载阀。进气阀在缸内压力稍小于进口压力时开启。排气阀在缸内压力稍大于排气压力时开启。

图 22-1(a)所示为气缸内充满空气的基本单元。在理论压容图(指示图)上,点 1 是压缩的起始点,两个阀门都关闭。

图 22-1(b)所示为压缩冲程,活塞运动到左端之后,原来的空气体积减小,而伴随着压力升高。阀门保持关闭。压容图表示从点 1 到点 2 的压缩,并且气缸内的压力已

达到所需的压力。

图22-1(c)所示为活塞完成了排气冲程。排气阀恰好在点2后打开。被压缩的空气通过排气阀流到容器内。

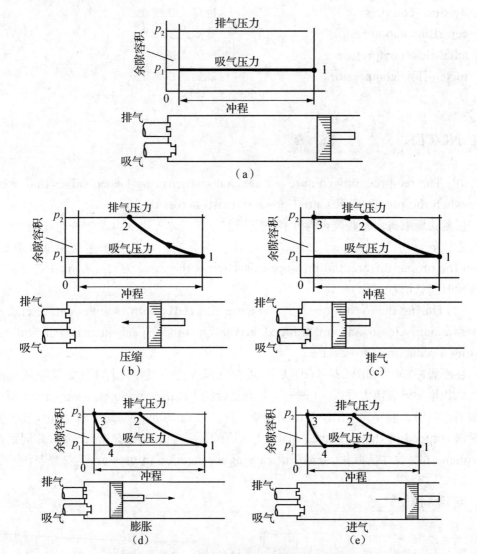

图22-1　往复式压缩机循环的各个步骤

活塞到达点3后,排气阀将关闭留下充满排出压力下空气的余隙空间。在膨胀冲程中,如图22-1(d)所示,进排气阀保持关闭并且封闭在余隙空间的空气体积增加导致压力降低。该过程随活塞向右运动而继续,直到气缸内压力降低到低于进气压力的点4。这时进气阀将开启,而空气将流进气缸直到在点1反向冲程的终端。这是进气或吸气冲程,如图22-1(e)所示。在压容图上的点1,进气阀将关闭而在曲轴的下一次循环周期重复。

在简单的两级往复式压缩机中,气缸与总压缩比成正比。第二级气缸因气体已部分被压缩和冷却而较小,气体占据的体积小于第一级进口处的体积。

请看压容图(如图22-2所示),第一级和第二级的起始压缩条件分别是点1和点5,压缩后为点2和点6,而在排气阶段后为点3和点7。随着活塞从点4和点8反向移动,封闭在余隙空间中的气体膨胀,在吸气冲程中在点1和点5气缸又重新充满气体,又进行重复循环。

任何容积式多级压缩机都遵循以上方式。

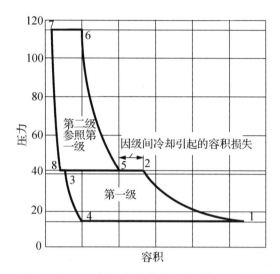

图22-2　双级、双缸100磅/平方英寸容积式压缩机组合式理论指示图

References

［1］卜玉坤.大学专业英语——机械英语 1［M］.北京:外语教学与研究出版社,2015.

［2］常红梅.数控技术应用专业英语［M］.2 版.北京:化学工业出版社,2012.

［3］程安宁,周新建.机械工程科技英语［M］.西安:电子科技大学出版社,2007.

［4］董建国.机械专业英语［M］.2 版.西安:电子科技大学出版社,2015.

［5］胡占齐,董长双,常兴.数控技术(NUMERICAL CONTROL TECHNOLOGY)［M］.武汉:武汉理工大学出版社,2004.

［6］《机电英语》教材组.机电英语［M］.2 版.北京:高等教育出版社,2007.

［7］李庆芬.机电工程专业英语［M］.哈尔滨:哈尔滨工程大学出版社,2010.

［8］李玉萍.机电英语［M］.2 版.北京:北京大学出版社,2021.

［9］刘建雄,王家惠.模具设计与制造专业英语［M］.北京:北京大学出版社,2006.

［10］刘杰辉.机械专业英语阅读教程［M］.大连:大连理工大学出版社,2005.

［11］刘瑛,罗学科.数控技术应用专业英语［M］.2 版.北京:高等教育出版社,2011.

［12］马佐贤.数控技术专业英语［M］.3 版.北京:北京理工大学出版社,2016.

［13］上海市职业技术教育课程改革与教材建设委员会.机电与数控专业英语［M］.北京:机械工业出版社,2012.

［14］施平.机电工程专业英语［M］.哈尔滨:哈尔滨工业大学出版社,2017.

［15］宋永刚.工程机械专业英语［M］.北京:人民交通出版社,2006.

［16］王晓江.模具设计与制造专业英语［M］.3 版修订版.北京:机械工业出版社,2021.

［17］徐鸿,董其伍.过程装备与控制工程专业英语［M］.北京:化学工业出版社,2018.

［18］杨晓红,王浩钢.模具专业英语［M］.北京:人民邮电出版社,2015.

［19］张黎明.机电工程专业英语［M］.北京:化学工业出版社,2010.

［20］张明文.工业机器人专业英语［M］.武汉:华中科技大学出版社,2017.

［21］ANTON R,KARI T,GALINA D,HASSAN H. Mechatronics technology and transportation sustainability［J］. Sustainability,2018,19(2):34-38.

［22］KEVIN Y,KRZYSZTOF I. Technologies for smart sensors and sensor fusion［M］. Boca Raton:CRC Press,2017.

［23］SHRABANI S,SASMITA B. A review on modern control applications in wind energy conversion system［J］. Energy & Environment,2019,14(2):12-23.

［24］WILAMOWSKI B,IRWIN D. Control and mechatronics［M］. Boca Raton:CRC Press,2018.